Recent Advances and Allied Applications of MXenes

Edited by

Inamuddin[1], Mohammad Abu Jafar Mazumd[2,3],
Mohamamd Luqman[4], Mohammad Faraz Ahmer[5]

[1]Department of Applied Chemistry, Zakir Husain College of Engineering and Technology, Faculty of Engineering and Technology, Aligarh Muslim University, Aligarh-202002, India

[2]Chemistry Department, King Fahd University of Petroleum & Minerals, Dhahran 31261, Saudi Arabia

[3]Interdisciplinary Research Center for Advanced Materials, King Fahd University of Petroleum & Minerals, Dhahran 31261, Saudi Arabia

[4]Department of Chemical Engineering, College of Engineering, Taibah University, Yanbu, Saudi Arabia

[5]Department of Electrical Engineering, Mewat Engineering College, Nuh Haryana, India

Published by **Materials Research Forum LLC**
Millersville, PA 17551, USA

Published as part of the book series
Materials Research Foundations
Volume 155 (2024)
ISSN 2471-8890 (Print)
ISSN 2471-8904 (Online)

Print ISBN 978-1-64490-286-8
eBook ISBN 978-1-64490-287-5

This book contains information obtained from authentic and highly regarded sources. Reasonable efforts have been made to publish reliable data and information, but the author and publisher cannot assume responsibility for the validity of all materials or the consequences of their use. The authors and publishers have attempted to trace the copyright holders of all material reproduced in this publication and apologize to copyright holders if permission to publish in this form has not been obtained. If any copyright material has not been acknowledged please write and let us know so we may rectify this in any future reprints.

Distributed worldwide by

Materials Research Forum LLC
105 Springdale Lane
Millersville, PA 17551
USA
https://www.mrforum.com

Manufactured in the United States of America
10 9 8 7 6 5 4 3 2 1

Table of Contents

Preface

Since 2011, the MAX phase derivatives MXenes are constantly expanding in terms of crystallographic and composition variability. Several advancements have been made in the previous few years that have accelerated the production of novel MXenes with better chemical diversification and crystal structures, which is uncommon in other two-dimensional (2D) material families. Due to their diverse elemental composition options, surface functional tunability, varied magnetic sorting, and considerable spin-orbit interactions, MXenes can truly be called multifunctional materials. Moreover, MXenes have also gained interest for a variety of applications, including catalysts, ion batteries, gas storage media, and sensors; as per their wide surface areas, hydrophilicity, adsorption ability, and strong surface reactivity. It is essential to stay current on diverse qualities and applications, due to the rapid growth of MXene-based technological research.

This book is intended to present the latest applicational advancements in diversified sectors. The summaries of the chapters are given below:

Chapter 1 introduces MXene as a promising electroactive material for supercapacitor application. Various types of MXene-based supercapacitors such as symmetric, asymmetric, micro, and transparent supercapacitors are mainly discussed in this chapter. Different factors that affect the electrochemical performance of MXene are also discussed.

Chapter 2 discusses the 2D material MXene with EMI shielding capacity. Better EMI shielding mechanisms within 2D layers and enhanced absorption in various composites of MXene are summarized.

Chapter 3 provides a detailed discussion of the non-linear optical behavior of MXenes. Experimental and DFT simulation results about structural, electronic, and optical properties of various types of MXene have been discussed which emphasize potential implementation in optoelectronic devices such as photodetectors and LEDs.

Chapter 4 discusses the numerous synthesis techniques of MXenes like etching (HF etching), non-HF etching, and hydrothermal method, in brief. The excellent properties of MXenes such as mechanical, electronic, optical, and magnetic properties, are also discussed in detail. In addition, the use of MXenes as photodetectors is highlighted.

Chapter 5 discusses the application of MXenes as an electro-catalyst. MXenes and its doped composite catalyst have been summarized for hydrogen evolution reaction, oxygen reduction reaction, batteries, supercapacitor, nitrogen reduction reaction, and CO_2 reduction reaction. The synthesis, and properties of various composite catalysts along with their structure-property relationship are briefed.

Recent Advances and Allied Applications of MXenes
Materials Research Foundations 155 (2024) 1-27

Materials Research Forum LLC
https://doi.org/10.21741/9781644902875-1

Chapter 1

Applications of MXenes in Supercapacitors

Arishma Buragohain, Debajyoti Mahanta*

Department of Chemistry, Gauhati University, Guwahati, Assam, India

debam@gauhati.ac.in

Abstract

In search of eco-friendly, low-cost, highly efficient energy materials to fulfil the global energy need, the research community from all over the world has explored a novel two-dimensional material called MXene. The term 'MXene' refers to a class of materials with very unique features including large surface area, high conductivity, etc., that could be used as the electrode material to fabricate highly efficient supercapacitors with high power density. Moreover, the composites made of MXene with some metal oxides, conducting polymers, and carbon-based materials exhibit very good electrochemical properties as electrode materials of supercapacitors.

Keywords

MAX Phase, MXene, MXene-Based Composite, MXene Symmetric Supercapacitor, MXene Asymmetric Supercapacitor, MXene Micro Supercapacitor, MXene Transparent Supercapacitor

Contents

1. Introduction

The present world in which we are living completely depends on energy. It will become nearly impossible to meet the global energy need by using conventional energy sources. The term conventional energy source, refers to natural gas, petroleum, coal, nuclear energy sources, etc. The United State Energy Information Administration (EIA) estimates that around 85% of global energy consumption relies on conventional sources of energy (petroleum, natural gas, coal, nuclear energy) as well as other liquids. Renewable sources of energy such as biomass/ biogas, hydropower, solar, geothermal, wind, tidal, and others account for only the remaining 15% of global energy consumption. The majority of energy used in the world is generated from conventional resources. However, the scarcity of conventional energy resources and their environmental impact associated with greenhouse gas like CO_2 release will cause a serious energy crisis in the future. Although renewable energy has a high potential to replace traditional energy, its regular production and efficient storage is still challenging.

About 580 million terajoules (1 terajoule= 10^{12} joule) of energy is consumed globally in a year, equivalent to the energy produced from 13,865 million tons of oil burning. The demand for global energy consumption may be increased to 740 million terajoules by the year 2040. Therefore, the development of a highly efficient and sustainable energy storage

Recent Advances and Allied Applications of MXenes   Materials Research Forum LLC
Materials Research Foundations 155 (2024) 1-27   https://doi.org/10.21741/9781644902875-1

technique becomes a global challenge for the research community to meet this large and increasing energy demand of approx. 7.9 billion world total population.

The energy storage technique can be defined as storing any form of energy for the future to reduce the gap between production and supply. Energy storage techniques include-(a) Electrochemical (b) Magnetic (c) Pneumatic (d) Thermal (e) Mechanical, and (f) Hydro systems [1]. Here in this chapter, we will only discuss electrochemical energy storage systems, the remaining systems are beyond our limit. Electrochemical energy storage techniques include batteries, fuel cells, capacitors, supercapacitors (SCs), etc. Among these, the SC is considered one of the most interesting storage devices for electrochemical energy.

Supercapacitors and batteries are the dominant storage devices for electrochemical energy that are commercialized on the global market. Li-ion batteries and Pb-ion batteries are used in most applications nowadays, there are some serious problems associated with batteries, although they have a high energy density. A supercapacitor is defined as a storage device of electrochemical energy that stores charges in the interface layer of the electrode and electrolyte of the system i.e., at the electrical double layer. Supercapacitors provide higher power density than conventional batteries such as Li-ion batteries. They have prolonged cycle life and lesser charge-discharge time as compared to the battery [2]. Moreover, SCs are quite less hazardous than batteries which gives an assurance towards the safety issues. They have also some other advantages such as a wide thermal range, lightweight, low maintenance, etc. Two similar or dissimilar electrodes, a suitable electrolyte, and a separator between the electrodes are the three fundamental components of SC. Mainly, the material used to fabricate electrodes for SC is a very crucial element for the efficacy of the device.

As shown in Fig.1, SC can be classified into basically three categories based on the charge storage mechanism. These are (1) Electrochemical Double Layer Capacitors abbreviated as EDLC, (2) Pseudo Capacitor, and (3) Hybrid Supercapacitor or Asymmetric Supercapacitors.

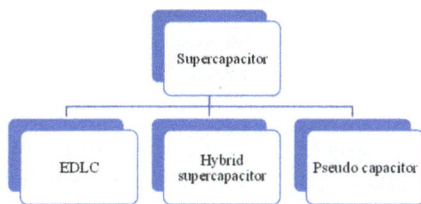

Fig.1: Classification of SCs [3].

In the EDLC, two identical carbon-based electrodes are arranged symmetrically and its electrode-electrolyte interface keeps the charges in a capacitive manner. Generally, commercial SCs are EDLC type and the charge storage mechanism is dependent on the reversible electrosorption of the ions at the interface of the electrode and electrolyte. In EDLC, no oxidation or reduction reaction happens and the storage technique for charges is non-faradaic. Again, the accumulation of the charges on the electrode surface of EDLC happens when a potential is applied and forms a layer of charges. It will again attract the oppositely charged ions and form another layer of ions. As a result, diffusion of ions in electrolytes occurs [3]. In contrast, pseudo capacitor stores charge in a faradic manner [2]. Pseudo capacitor electrodes are fabricated with various metal oxide, metal-doped carbon, or conducting polymers. When a potential is applied through the system, the oxidation-reduction reaction will occur in the pseudo capacitor electrode surfaces. The passages of charges occur across the double layer of the cell, which results in a faradic current. On the other hand, hybrid SCs are a combination of both the capacitor-type electrode as a source of power and the battery-type electrode as a source of energy [4].

The electrical properties such as the capacitance of the SCs are greatly influenced by the intrinsic properties of the electrode materials used. Various carbon materials (graphene, nanostructured carbon, activated carbon, carbon aerogel, etc.), oxides of metal (MnO_2, RuO_2, etc.), conducting polymer (polypyrrole, polyaniline, etc.), hybrid materials (polyaniline coated graphene oxide), etc. are used as electrode materials in SCs [5].

The development of highly efficient materials with a large surface area for electrodes to construct high-power devices is still a research hotspot in the field of energy. MXene, a 2D layered material, opens up a new exciting area as electrode material for energy storage devices especially for SCs as a consequence of its graphene-like layered structure, high electronic conductivity, hydrophilicity, and very rich surface chemistry.

2. Brief idea of MAX phase and MXene

A very unique ceramic material called MAX phase was discovered in 1960 which is a revolutionary step toward the field of advanced material. The newly discovered material MAX phase has some interesting characteristics such as high stiffness, lightweight, machinability, low-cost synthesis precursor, resistance to oxidation, resistance to thermal degradation, etc. [6].

The term 'MAX phase' can be explained by the common formula, $M_{n+1}AX_n$ where 'M' represents the d- block elements V, Cr, Ti, Y, Mo, Nb, Zr, Ta, Hf, and W. Similarly, 'A' corresponds to the p-block element like Al, Ga, In, Si, Ge, Sn, P, As, Se, S, Pb, Tl, which belong to 13,14,15, and 16 groups in the periodic table, 'X' represents C, N, B and 'n' is a natural number, viz. 1, 2, 3, 4 or 5 [6,7]. MAX phase therefore can be defined as the 3D layered ternary carbides or nitrides [8], and carbonitrides [9]. Up to now, over seventy MAX phases have been discovered by various researchers from all over the world [10]. Depending on the number of atoms of the M, A, and X elements in a particular MAX phase, they are categorized as 211(M=2, A=1, X=1), 312(M=3, A=1, X=2), 413(M=4, A=1, X=3) and 514(M=5, A=1, X=4). The example of 211(M_2AX) are Ti_2AlC, Nb_2AlC, Zr_2SnC, Hf_2SnC, Zr_2PbC, Ti_2AlN, Cr_2AlC, Ta_2AlC, V_2AlC, V_2PC, Nb_2PC, Ti_2PbC, Hf_2PbC, Zr_2SC, Nb_2SC, Hf_2SC, Zr_2AlC etc. Similarly, Ti_3AlC_2, Ti_3GeC_2, Ti_3SiC_2, Zr_3AlC_2, Hf_3AlC_2 etc. are example of 312 (M_3AX_2) type MAX phase. Ti_4AlN_3 and Ta_4AlC_3 are the 413 (M_4AX_3) type and Mo_4VAlC_4 is the example of the 514 (M_5AX_4) type MAX phase [8,9,11,12].

If we selectively etched out the element 'A' from the precursor MAX phase, a new class of two-dimensional layered material similar to graphene structure is formed which is called MXene (pronounced as 'Maxine'). In the year of 2011, Gogotshi, Barsoum, and co-workers at Drexel University synthesized MXene for the first time [13]. Although more than 70 different MXenes are theoretically predicted, only around 25 MXenes have been made since 2011. MXenes are described as $M_{n+1}X_nT_x$, where 'M' represents Ti, Zr, Hf, Y, V, Nb, Ta, Cr, Mo, W, etc. transition metals, the elements on the blue background in Fig.2; similarly, 'X' represents C, N as well as B, the elements on the grey background in Fig.2 and 'n' represents natural number 1, 2, 3, 4 or 5 [9,14]. Again, 'T_x' is the functional group -F, -Cl, -OH, or -O present on the surface of the $M_{n+1}X_n$ layer which is introduced throughout the etching procedure.

In MXene, two, three, four, or five layers of metal are interleaved with the X element. Therefore, we can get M_2XT_x, $M_3X_2T_x$, $M_4X_3T_x$, $M_5X_4T_x$, etc. different classes of MXene with one transition metal in the M site. The arrangement of elements M, X, and T in M_2XT_x, $M_3X_2T_x$, and $M_4X_3T_x$ MXenes are schematically shown in Fig.2.

However, more than one type of transition metal atoms can occupy the M site of MXene providing some new classes of MXene with different compositions namely "out-of-plane ordered MXene", "in-plane ordered double transition metal MXene", "in-plane ordered vacancy MXene" and "solid solution on M site MXene". The "out of plane ordered double transition metal MXene" possesses the general formula $M^I_2M^{II}X_2T_x$ and $M^I_2M^{II}_2X_3T_x$, where M^I and M^{II} represent two different transition metals. Again, the composition of "in-plane ordered double transition metal MXene" is $(M^I_{2/3}M^{II}_{1/3})_2XT_x$, and that of "in-plane ordered vacancy MXene" is $M^I_{4/3}XT_x$. Similarly, four different compositions of solid solution on M site MXene are found, and these are $(M^I_yM^{II}_{1-y})_2XT_x$, $(M^I_yM^{II}_{1-y})_3X_2T_x$, $(M^I_yM^{II}_{1-y})_4X_3T_x$, and $(M^I_aM^{II}_bM^{III}_cM^{IV}_{1-a-b-c})_4X_3T_x$. Moreover, solid solutions on X site MXene with compositions $M_2(C_yN_{1-y})T_x$ and $M_3(C_yN_{1-y})_2T_x$ are also found [15].

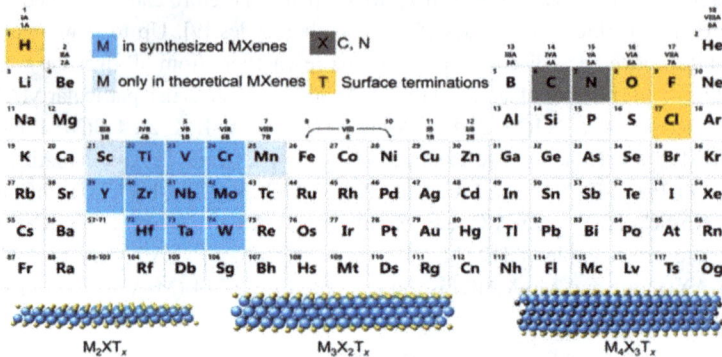

Fig.2: Periodic table showing the M, X, as well as T elements used to construct MXene ($M_{n+1}X_nT_x$) and Schematic for M_2XT_x, $M_3X_2T_x$, and $M_4X_3T_x$ MXene structure. Reprinted with permission from Ref. [14], Copyright (2022) American Chemical Society.

The quality of MXene greatly depends on the synthesis strategy adopted. Therefore, various preparation strategies for MXene have been developed. Generally, MXene is prepared from MAX phase precursor by HF etching, acid/ fluoride salt, or hydro fluoride etching, where $(NH_4)HF_2$, HCl/NaF, HCl/FeF_3, HCl/NH_4F, HCl/LiF, and HCl/ KF, etc. fluoride containing salts are used. Moreover, there are other methods also available for the synthesis of MXene; these are alkali etching, molten salt etching, electrochemical etching, chemical vapor deposition, hydrothermal or solvothermal method, self-assembly, etc. Fig. 3 shows the etching of the MAX phase and their exfoliation schematically along with the scanning electron micrograph for Ti_3AlC_2 MAX phase in Fig.3(A) in addition to $Ti_3C_2T_x$ MXene in Fig.3(B). The transformation from a three-dimensional MAX phase to a two-dimensional layered MXene structure is observed in these SEM images. Additionally, the transmission electron micrograph in Fig. 3(C) shows nice overlapping of the single layers of $Ti_3C_2T_x$ MXene.

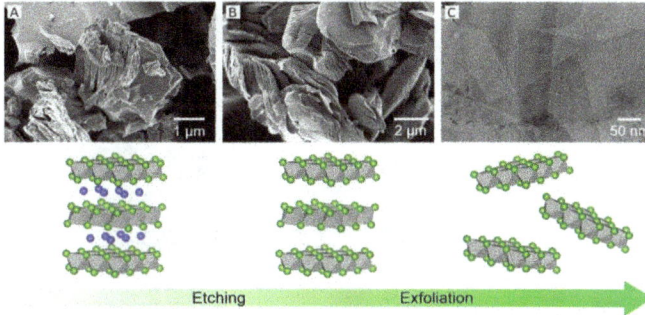

Fig.3: Diagrammatic representation of the MAX phase etching and their exfoliation to obtain 2D layered MXene. (A) scanning electron micrograph of Ti_3AlC_2 MAX phase. (B) scanning electron micrograph of $Ti_3C_2T_x$ MXene after selective etching of Al layer from MAX phase. (C) transmission electron micrograph of $Ti_3C_2T_x$ MXene where the overlapping of the single MXene layers is observed. Reprinted from ref. [16], copyright (2021), with permission from Elsevier.

All the MXene synthesis strategies can be categorized into two types namely the top-down method and the bottom-up method as shown in Fig.4 [17]. In the top-down strategy, single-layered MXene is synthesized by the 'A' layer etching selectively from the MAX phase. In contrast, the bottom-up strategy involves the growth of MXene from molecules or atoms [18]. The top-down approach of MXene synthesis is preferable because of the higher scale and lower cost of production as compared to the bottom-up approach.

These novel 2D advanced MXene materials have some very useful properties. These are good electrical conductivity, great stability, large specific surface area, optical properties, enriched surface mechanism due to excellent functional ability, highly dispersive in water-like solvents, environmentally friendly in nature, hydrophilicity, mechanical properties, etc. [18,19]. MXene can be categorized as strong material yet they are bendable. The in-plane Young's modulus values for Ti_3C_2, Ti_2C, and Ti_4C_3 MXene are 502, 597, and 534 GPa respectively. MXene possesses a very large electrical conductivity of greater than 10,000 Scm^{-1}. They possess negative zeta potential values from -30 to -80 mV. This negative value of zeta potential suggests that high weight percent (up to 70%) colloidal suspension of MXene can be easily made without any surfactant. Also, in polar-aqueous solvent colloidal stability of MXene is high. Because of these amazing chemical and physical properties, they are employed in various applications such as separation techniques, photocatalysis, and adsorption [20]. In addition, they can be employed in storage devices of electrochemical energy mainly in SCs as electrode materials.

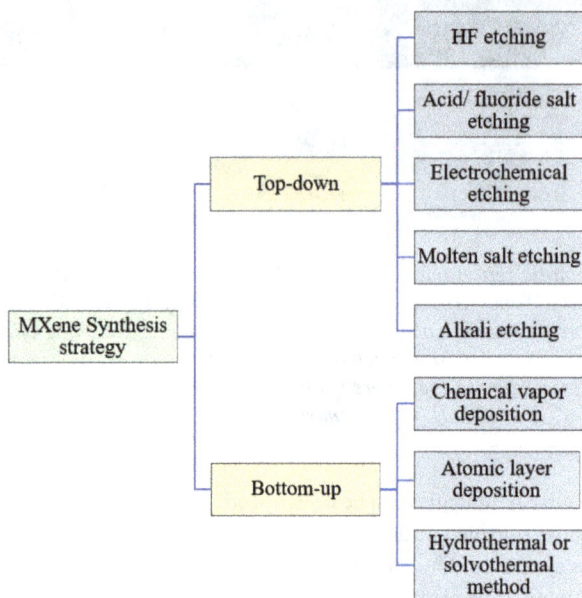

Fig.4: Different strategies for MXene synthesis [17].

3.   MXene and MXene-based composites as supercapacitor electrode materials

Due to 2D graphene-like structure, excellent electrical conductivity, extraordinary environmental stability, surface functionality, and hydrophilic nature, MXenes have been explored as excellent electrode materials for SC-like energy storage devices. Although around 25 different MXenes have been reported, less than half of them (~10) are used as electrode materials in supercapacitors so far. These are $Ti_3C_2T_x$, Ti_2CT_x, V_2C, V_4C_3, $Nb_4C_3T_x$, $Mo_{1.33}CT_z$, $Mo_2TiC_2T_x$, etc. [21–25]. Not only MXene but MXene-based composites are also utilized as materials for supercapacitor electrodes. In the MXene-based composite, other materials are introduced to improve the electrochemical performance of MXene. Handling and flexibility are also improved along with the electrochemical performance in MXene composite materials as compared to parent MXene. Following are some examples of MXene based composite materials used in various supercapacitors-

$Ti_3C_2T_x$ MXene-reduced graphene oxide composite, $Ti_3C_2T_x$/ polypyrrole, $Ti_3C_2T_x$ @ polydiallyldimethylammonium chloride (PDDA), Ti_3C_2/ ZnO, $Ti_3C_2T_x$/ polyaniline, $Mo_{1.33}C$/Poly(3,4-ethylene dioxythiophene):Polystyrene sulfonate, Ti_3C_2/ WS_2, Polyaniline@TiO_2/ $Ti_3C_2T_x$ ternary composite, $Ti_3C_2T_x$/ MoO_3 composite, MXene/ Layered metallic double hydroxide, $Ti_3C_2T_x$/ MnO_2 nanowire, Ti_3C_2/ CuS composite, $Ti_3C_2T_x$/ Co-MOF composite, $Ti_3C_2T_x$/ α-Fe_2O_3 nanocomposite, Ti_3C_2/ FeOOH quantum dots, Ti_3C_2/ TiO_2 nanocomposite, $Ti_3C_2T_x$/ carbon nanotube, and $Mo_{1.33}CT_z$/ cellulose composite [22–42].

4.   Parameters that affect the electrochemical behaviors of MXene

Several parameters greatly affect the electrochemical performance of MXene as well as its composition as SC electrode materials. The following are the factors that are identified –

4.1   Etchant

Different types of etchants are used for selective etching of the layer containing element 'A' from the MAX phase to make MXene. Aqueous HF or adding a fluoride-containing salt such as LiF, NH_4F, NaF, KF, FeF_3, etc. in hydrochloric acid for the in-situ generation of HF are applied as the etchant. The ions which will be intercalated between the two sheets of MXene depend on the type of etchant used. Depending on the ions intercalated, the restacking procedure of MXene sheets is prevented during the synthesis process. Restocking will again affect the electrochemical parameter such as volumetric and gravimetric capacitance. For example, we can compare the etching of the MAX phase by using HF and HCl/LiF etchant which is a milder etchant than HF. If the etchant is HF, there is a possibility of re-assembling of single sheets of MXene because of the smaller size of the H^+ ion. But the availability of interlayer spacing can be increased by using HCl/LiF etchant. During the synthesis procedure, the lithium-ion (Li^+) and the water molecule prevent the restacking of the single delaminated sheets of MXene. Thus, the interlayer spacing can be tuned by using different etchants [17,43].

4.2   Etchant concentration

Etchant concentration is one of the most crucial factors that influence MXene's electrochemical performance as an electroactive material. One can adjust the functionality of the MXene surface by changing the concentration of the etchant which will greatly affect the electrochemical performance as the electrode material of the SC. In a low concentration of etchant, there will be a large number of H_2O molecules that can accommodate as well as expand the interlayer space compared to the higher concentration of etchant. In the case of HF etching, lower concentration always means a higher -O to -F functional group ratio.

More hydrogen ions will be accessible when more H_2O molecules are intercalated within the MXene sheets. Well-separated MXene with a large number of intercalated water molecules provides better electrochemical performance such as high capacitance as well as long cyclic stability. If we compare the electrochemical behavior of $Ti_3C_2T_x$ prepared from the Ti_3AlC_2 by using two different concentrations of HF namely 6M and 12M, then it was observed that a more symmetrical and rectangular-shaped cyclic voltammogram as shown in Fig.5(a) was obtained for $Ti_3C_2T_x$-6M which suggests the better capacitive performance of $Ti_3C_2T_x$-6M than that of $Ti_3C_2T_x$-12M electrode. Moreover, the capacitance of $Ti_3C_2T_x$-6M was 129 Fg^{-1} higher than that of $Ti_3C_2T_x$-12M {Fig.5(b)}.

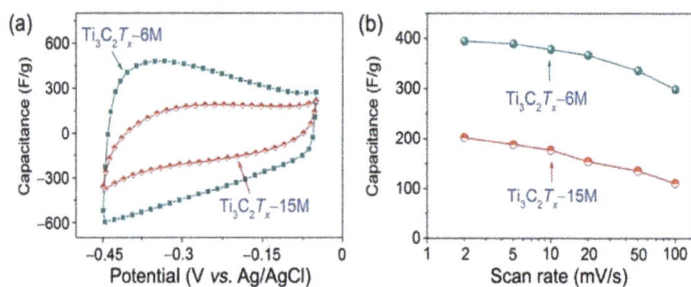

Fig.5: Capacitive behavior of $Ti_3C_2T_x$-6M ($Ti_3C_2T_x$ prepared with 6M HF solution) and $Ti_3C_2T_x$–15M ($Ti_3C_2T_x$ prepared with 15M HF solution) electrodes in 1M H_2SO_4 electrolyte. (a)Capacitance vs potential graph of $Ti_3C_2T_x$-6M and $Ti_3C_2T_x$–15M (scan rate 20 mVs^{-1}). (b) Comparison of the gravimetric capacitance of both the electrodes (at scan rates 2-100 mVs^{-1}). Reprinted with permission from ref. [44], Copyright (2022) American Chemical Society.

Moreover, the charge transfer resistance (R_{ct}) measured by the electrochemical Impedance Spectroscopy is also lower for the MXene which is synthesized using a low concentration of etchant. It means that MXene with a low concentration of etchant acquires a larger electro-active surface area to make good contact with H_3O^+ ions [44].

4.3 Surface termination group

Any of the anions, e.g., Cl^-, Br^-, I^-, PO_4^{3-} or SO_4^{2-} can be inserted in the MXene sheets as a surface functional group using the corresponding acid along with HF etchant in the synthesis process of MXene. Although these anions are not found intercalated within the MXene sheets, they still play a salient role in the swelling along with the drying of MXene sheets. They also impact the intercalation and deintercalation behavior of MXene. These anions are preferentially absorbed on the edge of the MXene and help to enlarge the interlayer spacing as a consequence of their sizable-ionic radius. This will greatly enhance the ion diffusion process leading to high volumetric capacitance or/and gravimetric capacitance of the prepared SC device [17,45]. Kajiyama et al. reported that the presence of the -Cl termination group in the $Ti_3C_2T_x$ MXene results in an unprecedentedly high volumetric capacitance value of 130 Fcm^{-3} and gravimetric capacitance of 300 Fg^{-1} in Li-ion hybrid capacitors. It was said that Li-ion accessibility increases due to the expanded presence of steric Cl^- ion [46].

4.4 Partial etching of 'A' group from the MAX phase

The problem of restacking the MXene sheet as well as the low conductivity can be sorted out by partial etching of the 'A' group from the structure of MAX phase during the etching process. A large space will remain open for ion diffusion in case of partial etching of aluminum from the Ti_3AlC_2. Moreover, the remaining Al works as an 'electron bridge'. As a result, conductivity between the MXene sheet is greatly enhanced, and so does the capacitance [17].

4.5 Etching time and etching temperature

The consequences of etching duration and temperature on the electrochemical performance of the MXene are not remarkable unlike other factors [47].

5. Different types of supercapacitors with MXene

5.1 MXene-based symmetric supercapacitor

Symmetric supercapacitors (SSCs) are those SCs that consist of two identical electrodes with a suitable electrolyte. MXene-based materials are used to construct various SSCs and they can be categorized as one-dimensional, two-dimensional, and three-dimensional SCs according to their dimensionality.

5.1.1 One-dimensional (1D) supercapacitor

The SC constructed with one-dimensional fiber-based materials, such as polymer fiber, metal fiber, etc. is called as 1D SC device. The example of polymer fibers that are used to make this kind of SC device are silver-plated nylon fiber, Polyester yarn, and the example of metal fiber is stainless steel wire [48–50]. The advantage of using a wire-like current collector is the easy electrode fabrication as these wires could be knitted as electronic textiles by using well-developed textile technology. The chosen fiber should possess good electrical conductivity and it should be mechanically enough tough so that it can be woven in practical production. They should be commercially available and of low cost. This type of flexible fiber-based SC may be a promising candidate for the making of wearable as well as portable smart electronics. The flexible nature of the fiber allows the prepared MXene-coated fiber electrode to deform in any complex shape, which makes them very special. The capacity retention of the $Ti_3C_2T_x$ MXene-coated nylon fiber electrode continues to exist beyond 90% on bending or twisting. Again, the capacitance ratio, C/C_0, where C stands for the capacitance before and C_0 the capacitance after bending or knotting remains 82% even after knotting the fiber electrode. No changes in the cyclic voltammogram and electrochemical impedance spectra occur even after 100° bending, twisting, and knotting of the MXene-coated silver-plated nylon fiber-based SC. However, a negligible reduction in the specific capacitance is noted after deformation and only a few defective interfaces are also observed after 100° bending. All these features reveal that one-dimensional $Ti_3C_2T_x$ MXene coated nylon fiber SC provides excellent flexibility, knittability along with excellent cyclability [48]. Moreover, fibrous substrates enhance the mechanical strength of the devices. To prepare this type of electrode for MXene-based 1D SC, various methods are employed. Some of these are electrospinning, brush coating, and solution-processed methods [48-50], etc. In the electrospinning method, polyacrylic acid, polyethylene oxide, polyvinyl alcohol, etc. polymers are used to prepare an electrospinning solution with MXene [51]. The brush coating method is a simple and easy method. Shao et al. prepared PET@ MXene nanofibre-coated yarn via electrospinning which shows all the necessary properties such as strength, flexibility, machinability, etc. The SSC device fabricated with PET@MXene nanofibre coated yarn electrode provides very high-power density, high specific capacitance of 18.39 mFcm^{-1} at 5 mVs^{-1} and 98.2% retention of specific capacitance after performing six thousand cycles. The capacitance ratio (%) of this electrode remained nearly unchanged after deformation as appeared in Fig.6 [49].

Although MXene-coated fibre electrodes provide the benefit of mechanical strength and flexibility to the supercapacitor, there is an issue of imperfectness at the interface of MXene and the fiber substrate, which remains unsolved. A free-standing MXene electrode is a probable solution to this kind of problem, which can be explored with a system such as $Ti_3C_2T_x$ MXene embedded in carbon nanofibre. To prepare $Ti_3C_2T_x$/carbon nanofiber, first $Ti_3C_2T_x$/polyacrylonitrile nanofiber was prepared by electrospinning technique followed by carbonization of polyacrylonitrile to prepare $Ti_3C_2T_x$/carbon nanofibre [52].

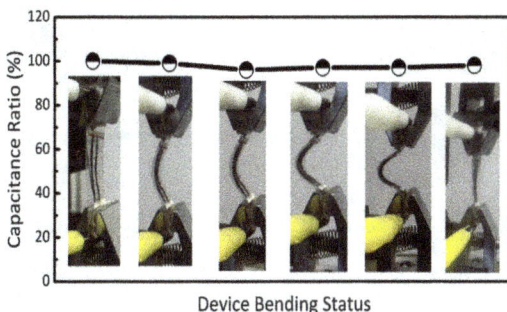

Fig.6: Capacitance ratio (C/C_o in percentage, where C_o represents the initial capacitance of the device) of PET@MXene nanofiber coated yarn SC under different bending stages and twisting (the last image in the graph). Reprinted from ref. [49], Copyright (2022), with permission from Elsevier.

5.1.2 Two-dimensional (2D) supercapacitor

MXene coated on a planar substrate, free-standing MXene films may be promising electrode materials for the fabrication of 2D SC devices. Highly conducting $Ti_3C_2T_x$ film, MXene/graphene, and $MXene/MnO_2$ electrodes can be utilized directly as free-standing MXene films for 2D SCs. Vacuum-assisted filtration technique is used to obtain this type of homogeneous film where MXene ink is filtrated in a vacuum [53,54]. Again, two-dimensional substrates such as commercially available A4 paper, textiles, polypropylene membranes, plastic sheets, etc. are also used to coat MXene. The A4 paper was first coated with Au keeping the thickness of about 200 nm by magnetron sputtering technique to provide conductivity to the paper. Again, the MXene paste prepared by mixing MXene, carbon black, and binder polyvinylidene fluoride (PVDF) in a suitable solvent was coated manually on the surface of Au-coated paper by using Meyer rod coating. Meyer rod coating is a manual technique to spread a paste over a two-dimensional planar surface with a rod as observed in Fig. 7(a). The direct laser machining technique is used to fabricate MXene/Au/paper electrodes over a large area. The laser beam was directly employed to slice the whole stack comprising of MXene/Au/paper electrode [55]. Since MXene is hydrophilic, it can be easily converted to MXene ink. As depicted in Fig.7, MXene ink is spin-coated, stamped, dip-coated, written, and printed over the planar substrate to make two-dimensional SC electrodes.

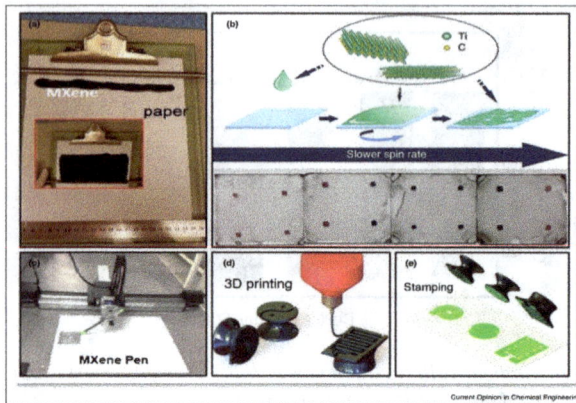

Fig.7: Various examples of electrode fabrication techniques using MXene inks- (a) Meyer rod coating (b) Spin coating, (c) Writing process, (d) 3D printing, and (e) Stamping technique [56].

5.1.3 Three-dimensional (3D) supercapacitor

SCs with a three-dimensional open porous structure provide faster ion transportation, high active mass loading, and larger interfacial area compared to compact 2D SCs, thereby enhancing the areal energy and power density. Li and co-workers designed a viscous pseudoplastic nanocomposite ink from $Ti_3C_2T_x$ MXene, Ag nanowire, MnO_2 nanowire, and fullerene to fabricate a micro-supercapacitor (MSC) device. They used extrusion-based 3D printing technology to print the MXene ink composite onto a substrate followed by unidirectional freezing which resulted in a 3D honeycomb-like microporous structure with a nacre-like lamella cell wall structure. The MXene sheets and manganese dioxide nanowire provide a high surface area internal structure which again gives high energy density per unit area, similarly, Ag nanowire gives rapid and efficient charge delivery throughout the bulk of the electrode. The presence of fullerene in the composite benefits as a lubricant to allow $Ti_3C_2T_x$ MXene layer slippage which provides the partial deformation to the honeycomb-like structure. This kind of MSC exhibits extraordinarily high areal capacitance, energy density, as well as power density, and also superior cyclability [57]. MXene sheet caging within the 3-dimensional tangled polypyrrole nanowires [58], MXene@ N-doped carbon foam [59], MXene @Ni-Al double layered hydroxide composite [30], MXene@ reduced graphene oxide composite aerogel [60] are the examples of electrode materials for MXene 3D SC device. In the case of the MXene@polypyrrole composite, the highly conductive and intrinsic redox active polypyrrole interwind on the surface of $Ti_3C_2T_x$ MXene sheet, giving a porous 3-

dimensional architecture with a huge number of voids. The presence of both the MXene sheet and the polypyrrole nanowire enhances the electrochemical performance including gravimetric capacitance, capacity retention, and cyclic stability of the SC device. Again, heteroatom-doped carbon foam can enhance the pseudocapacitance of the SC device and at the same time, it provides the structural support to the MXene sheet to form a 3D porous interconnected network-like structure [59].

Wang et al. fabricated a three-dimensional porous structure of an MXene composite by growing nickel-aluminum double-layered hydroxide platelet on an MXene sheet using the liquid-phase deposition method. The benefit of utilizing this type of supercapacitor is the excellent faradic utilization of the electroactive material [30].

5.2   MXene-based asymmetric supercapacitor

An asymmetric supercapacitor (ASC) has two dissimilar electrodes, where the cathode is a battery-type faradic electrode and the anode is an EDLC-type non-faradic electrode [61]. In a typical symmetric SC with an aqueous electrolyte, the potential window cannot be expanded beyond 1.23 V since above this potential limit the thermodynamic breakdown of water molecules happens. The potential window of a symmetric SC device fabricated with MXene is quite narrow which is about 0.6 V in aqueous electrolyte [62]. Beyond this voltage window, oxidation possibly happens to the MXene material. However, we can expand the potential window of the MXene SC and thus the energy density by fabricating an ASC by assembling it with a completely different electrode. The examples of different MXene ASCs are- $Ti_3C_2T_X$//CP@rGO, $Ti_3C_2T_X$//C_x@rGO, $Ti_3C_2T_X$//rGO, $Ti_3C_2T_X$//RuO_2, $Ti_3C_2T_X$//Mn_xO_n, $Ti_3C_2T_X$//MoS_{3-x}@3DnCF, etc. [7,62–66]. In $Ti_3C_2T_X$//CP@rGO, conducting polymers (CP) including polyaniline (PANI), poly 3,4-ethylene dioxythiophene (PEDOT), and polypyrrole (Ppy), coated on reduced graphene oxide (rGO) i.e., CP@rGO was used as the anode. This type of all-pseudocapacitive ASC can extend the voltage window as much as 1.45 V in a 3M H_2SO_4 electrolyte. Polyaniline-containing SC device i.e., $Ti_3C_2T_X$//PANI@rGO exhibited outstanding capacity retention of capacitance i.e., 88% after 20,000 cycles, and provided a very high energy density of 17 Whkg^{-1}, higher than any other polyaniline-containing and MXene- based devices [7]. Boota et al. Fabricated the '$Ti_3C_2T_X$//C_x@rGO' all-pseudocapacitive ASC device which is schematically represented in Fig. 8. In this device, $Ti_3C_2T_x$ MXene was used as the negative electrode, and C_x@rGO hybrid electrode was used as the positive electrode. In the C_x@rGO hybrid, C_x represents the quinone-functionalized viologen $(C_5H_4NR)_2^{n+}$ molecule which is deposited on rGO. This device can store its charges within the extended potential window of 1.5 V in a 1M sulfuric acid electrolyte. Moreover, the capacitive contribution was calculated as 85% (black shaded area in current vs voltage plot as demonstrated in Fig.

8) of the total charges stored in the device at 5 mVs^{-1} [63]. Similarly, in all the other above-mentioned examples of MXene-based ASCs, Ti$_3$C$_2$T$_X$ MXene electrodes were used as the negative electrode, and rGO, RuO$_2$, Mn$_x$O$_n$, MoS$_{3-x}$@3DnCF (3D printed nanocarbon framework), etc. were used as the positive electrode.

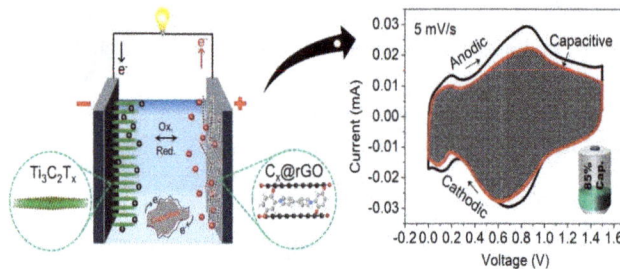

Fig. 8: Diagrammatic representation of Ti$_3$C$_2$T$_X$//C$_x$@rGO ASC and Current vs voltage plot showing the capacitive contribution (shaded region) and diffusion-limited contribution towards the total charge storage. The capacitive contribution is 85% of the total charge at 5 mVs^{-1} scan rate. Reprinted from ref. [63], copyright (2022), with permission from Elsevier.

There are some examples of ASCs where MXene-based composites are employed as the positive electrode. Two examples of such combinations are activated carbon//NiCo$_2$S@MXene [67], and graphene//graphene encapsulated MXene Ti$_3$C$_2$T$_X$@polyaniline ASCs [68].

5.3   Current MXene based micro-supercapacitor

Microsupercapacitor (MSC) can be defined as the miniatured SC ranging from centimetres to micrometres which can be integrated with electronic devices. Any symmetric or asymmetric SC can be converted into their miniature forms. The required footprint area i.e., the space for the micro-size devices is very small. Therefore, this type of SC device provides a great opportunity for the other functional devices to be compact in a small space in electronics, which leads to the development of microelectronics with higher working efficiency. In the earlier days, people used to prepare sandwich-type MSC by stacking the electrolyte in between the cathode and the anode films. It is an easy method but there is the possibility of short circuits, the displacement of the electrode films, and a lack of precise control over the electrode-electrolyte interlayers. Different types of MSC have been developed so far; for instance, sandwich-type MSCs, self-powered MSCs, parallel fiber

MSCs, twisted fiber MSCs, coaxial fiber MSCs, interdigital MSCs, and self-healable MSCs. The fabrication of the MSC generally follows a series of works starting from the patterning of the current collector followed by the coating of the electrode materials by spray coating, sputtering, electrophoretic/ electrochemical deposition, chemical vapor deposition, etc. [42,69]. Although various techniques have been developed for the fabrication of different types of MSCs, no one is dominating the field of microelectrode fabrication. These are photolithography, laser scribing, reactive ion etching, inkjet printing, screen printing, and so on [70].

Photolithography: The technique of photolithography or UV lithography is used in the precise as well as cost-effective fabrication of nano or micro-patterns over suitable substrates. This technique is used in the Micro-Electromechanical System (MEMS) as well as in the Nano-Electromechanical System (NEMS). These two are the techniques used in the designing of micro or nano-sized integrated devices that combine mechanical and electrical components. Photolithography is a very useful technique to prepare electrodes for MSC with a large number of active materials [31].

Scratch method: It is a direct scratching technique in which interdigitated electrode for a planar MSC is fabricated by simply scratching the film of electrode material using a common syringe needle. It is an easily achieved and almost zero-cost technique for patterning electrodes. Since it is a manual technique, various patterns can be achieved for planar MSC with different configurations such as parallel or series connections of two devices. Li et al. constructed a film of $Ti_3C_2T_x$ /graphene composite material onto SiO_x/Si wafer taking as substrate by using the direct scratch method [70].

Inkjet printing: Inkjet printing is again a powerful electrode fabrication technique in which a droplet of ink prepared from electroactive materials is pushed to reproduce a digital image in an interdigital pattern on diverse substrates such as paper and plastic. Solubility, chemical stability, surface tension, viscosity as well as particle size, etc. are the crucial properties of the ink which determine the quality of printing.

Screen printing method: In this method, the ink prepared from the electroactive material such as $RuO_2.xH_2O@Ti_3C_2T_x$MXene-Ag nanowire composite is spread over a screen mesh to create a pattern. To ensure the rheological properties of the prepared ink, the Ag nanowire is coordinated with $RuO_2.xH_2O@Ti_3C_2T_x$ MXene-based composite. Moreover, Ag nanowire provides a network-like structure that guarantees faster ion transfer [71]. The ink should have optimum viscosity as well as some suitable rheological properties for use in screen-printing techniques. This method gives in-plane, interdigital, flexible MSC.

Other methods are also available such as the Laser scribing method, the Reactive ion etching method, and so on.

Recent Advances and Allied Applications of MXenes   Materials Research Forum LLC
Materials Research Foundations 155 (2024) 1-27   https://doi.org/10.21741/9781644902875-1

5.4   MXene-based transparent supercapacitor

A monolayer of MXene is optically transparent in nature, e.g., the calculated transmittance values of a single sheet of $Ti_3C_2T_x$ MXene were 90.5% at 240 nm wavelength and 97.3% at 800 nm wavelength, which are quite close to the single layer of graphene [72]. Lower concentrated MXene ink can be used to fabricate electrodes for transparent SCs since the decay of transmittance occurs exponentially with the film thickness. The transparent MXene- based SCs electrodes are fabricated by using techniques like inkjet printing [73], spin-casting followed by vacuum annealing [74], spray coating [75], etc. The inkjet printing technique is challenging due to an effect called the 'coffee ring effect' in which most of the solute present in an ink is deposited on the outer edge of the printing pattern. Thus, the design of ink is an important task for this kind of printing technique. In addition to the solvent and solute size, the rheological properties of the MXene are also important parameters for a good quality ink [76].

Two simple figures of merits (FoM_e and FoM_c) are used to quantitatively evaluate the optoelectronic as well as the electrochemical performance of transparent electrodes. FoM_e can be described by the ratio σ_{DC}/σ_{opt} , where σ_{DC} and σ_{opt} stand for electrical and optical conductivity respectively. Hence, FoM_e corresponds to the optoelectronic performance of the transparent electrodes. On the other hand, FoM_c is defined by the volumetric capacitance, C_v to optical conductivity ratio (C_v:σ_{opt}) which specifies the electrochemical performance of the electrode. Quantitatively, FoM_e is obtained directly by putting the sheet resistance (R_s) and transmittance (T) value in the following equations-

$$T = (1 + \frac{Z_o \sigma_{opt}}{2R_s \sigma_{DC}})^{-2}$$

where Z_o represents the impedance of free space which is 377 Ω.

The sheet resistance of $Ti_3C_2T_x$ MXene film with 24% transmissivity was found to be 1.47 $\pm$ 0.1 kΩ sq^{-1} and areal capacitance 1225 μFcm^{-2} as given in Table-1 at 5 mVs^{-1} scan rate. This value of areal capacitance is quite good in comparison to any other transparent electrodes. Although $Ti_3C_2T_x$ film having 87% transmissivity acquires higher sheet resistance, it still provides reasonably good areal capacitance. The relationship between the optical and electrochemical properties of transparent SC devices can be established by these results. MXene film with 87 to 24% transmissivity corresponds to the figure of merit (σ_{DC}/σ_{opt}) of 0.0012 to 0.13 [76]. Three parameters namely capacitance, power density as well as both of the figures of merits of an electrode need to be maximized to construct an excellent transparent SC device [74].

Table-1. Areal capacitances of $Ti_3C_2T_x$ MXene film prepared by inkjet printing at two different transmissivities and sheet resistances [76].

Transmissivity	Sheet Resistance	Areal Capacitance
24 %	1.47 ± 0.1 kΩ sq^{-1}	1225 μFcm^{-2} @ 5 mVs^{-1}
87 %	1.66 ± 0.16 MΩ sq^{-1}	187.5 μFcm^{-2} @ 10 mVs^{-1}

Conclusion

SC is a promising storage device for electrochemical energy as a consequence of its high-power density, high cyclic stability, and environment-friendly nature. In search of energy materials for SC, MXene offers a viable option because of its large surface area and magnificent electrochemical properties. Here in this chapter, we have summarized a quick introduction to the MAX phase, MXene, and the application of MXene in different types SCs. Furthermore, there is an outline of composite materials of MXene that improves the electrochemical performance of SCs. The most frequently used MXenes in SC have either been $Ti_3C_2T_x$ or $Ti_3C_2T_x$-based composite materials. The electrochemical behaviors of most of the other MXene as SC electrode materials are still unknown.

References

[1] A.G. Olabi, Renewable energy and energy storage systems, Energy 136 (2017) 1-6. https://doi.org/10.1016/j.energy.2017.07.054

[2] A. Borenstein, O. Hanna, R. Attias, S. Luski, T. Brousse, D. Aurbach, Carbon-based composite materials for supercapacitor electrodes: A review, J. Mater. Chem. A. 5 (2017) 12653-12672. https://doi.org/10.1039/C7TA00863E

[3] Z.S. Iro, C. Subramani, S.S. Dash, A brief review on electrode materials for supercapacitor, Int. J. Electrochem. Sci. 11 (2016) 10628-10643. https://doi.org/10.20964/2016.12.50

[4] M. Vangari, T. Pryor, L. Jiang, Supercapacitors: Review of materials and fabrication methods, J. Energy Eng. 139 (2013) 72-79. https://doi.org/10.1061/(ASCE)EY.1943-7897.0000102

[5] P. Sharma, T.S. Bhatti, A review on electrochemical double-layer capacitors, Energy Convers. Manag. 51 (2010) 2901-2912. https://doi.org/10.1016/j.enconman.2010.06.031

[6] M. Barsoum, T.E. Raghy, The MAX Phases: Unique new carbide and nitride materials, Am. Sci. 89 (2001) 334. https://doi.org/10.1511/2001.28.334

[7] M. Boota, Y. Gogotsi, MXene-Conducting polymer asymmetric pseudocapacitors, Adv. Energy Mater. 9 (2019) 1-8. https://doi.org/10.1002/aenm.201802917

[8] T. Lapauw, K. Lambrinou, T. Cabioc'h, J. Halim, J. Lu, A. Pesach, O. Rivin, O. Ozeri, E.N. Caspi, L. Hultman, P. Eklund, J. Rosén, M.W. Barsoum, J. Vleugels, Synthesis of the new MAX phase Zr_2AlC, J. Eur. Ceram. Soc. 36 (2016) 1847-1853. https://doi.org/10.1016/j.jeurceramsoc.2016.02.044

[9] G. Deysher, C.E. Shuck, K. Hantanasirisakul, N.C. Frey, A.C. Foucher, K. Maleski, A. Sarycheva, V.B. Shenoy, E.A. Stach, B. Anasori, Y. Gogotsi, Synthesis of Mo_4VAlC_4 MAX phase and two-dimensional Mo_4VC_4 MXene with five atomic layers of transition metals, ACS Nano. 14 (2020) 204-217. https://doi.org/10.1021/acsnano.9b07708

[10] J. Chen, Q. Huang, H. Huang, L. Mao, M. Liu, X. Zhang, Y. Wei, Recent progress and advances in the environmental applications of MXene related materials, Nanoscale. 12 (2020) 3574-3592. https://doi.org/10.1039/C9NR08542D

[11] M. Roknuzzaman, M.A. Hadi, M.A. Ali, M.M. Hossain, N. Jahan, M.M. Uddin, J.A. Alarco, K. Ostrikov, First hafnium-based MAX phase in the 312 family, Hf_3AlC_2: A first-principles study, J. Alloys Compd. 727 (2017) 616-626. https://doi.org/10.1016/j.jallcom.2017.08.151

[12] M. Griseri, B. Tunca, T. Lapauw, S. Huang, L. Popescu, M.W. Barsoum, K. Lambrinou, J. Vleugels, Synthesis, properties and thermal decomposition of the Ta_4AlC_3 MAX phase, J. Eur. Ceram. Soc. 39 (2019) 2973-2981. https://doi.org/10.1016/j.jeurceramsoc.2019.04.021

[13] M. Naguib, M. Kurtoglu, V. Presser, J. Lu, J. Niu, M. Heon, L. Hultman, Y. Gogotsi, M.W. Barsoum, Two-dimensional nanocrystals produced by exfoliation of Ti_3AlC_2, Adv. Mater. 23 (2011) 4248-4253. https://doi.org/10.1002/adma.201102306

[14] Y. Gogotsi, B. Anasori, The rise of MXenes, ACS Nano. 13 (2019) 8491-8494. https://doi.org/10.1021/acsnano.9b06394

[15] M. Naguib, M.W. Barsoum, Y. Gogotsi, Ten years of progress in the synthesis and development of MXenes, Adv. Mater. 33 (2021) 1-10. https://doi.org/10.1002/adma.202103393

[16] L. Verger, C. Xu, V. Natu, H.M. Cheng, W. Ren, M.W. Barsoum, Overview of the synthesis of MXenes and other ultrathin 2D transition metal carbides and nitrides, Curr. Opin. Solid State Mater. Sci. 23 (2019) 149-163. https://doi.org/10.1016/j.cossms.2019.02.001

[17] M. Hu, H. Zhang, T. Hu, B. Fan, X. Wang, Z. Li, Emerging 2D MXenes for supercapacitors: Status, challenges and prospects, Chem. Soc. Rev. 49 (2020) 6666-6693. https://doi.org/10.1039/D0CS00175A

[18] J.A. Kumar, P. Prakash, T. Krithiga, D.J. Amarnath, J. Premkumar, N. Rajamohan, Y. Vasseghian, P. Saravanan, M. Rajasimman, Methods of synthesis, characteristics, and environmental applications of MXene: A comprehensive review, Chemosphere. 286 (2022) 131607. https://doi.org/10.1016/j.chemosphere.2021.131607

[19] L. Chen, J. Zhang, Q. Li, J. Vatamanu, X. Ji, T.P. Pollard, C. Cui, S. Hou, J. Chen, C. Yang, L. Ma, M.S. Ding, M. Garaga, S. Greenbaum, H.S. Lee, O. Borodin, K. Xu, C. Wang, A 63 m superconcentrated aqueous electrolyte for high-energy Li-ion batteries, ACS Energy Lett. 5 (2020) 968-974. https://doi.org/10.1021/acsenergylett.0c00348

[20] P. Kuang, J. Low, B. Cheng, J. Yu, J. Fan, MXene-based photocatalysts, J. Mater. Sci. Technol. 56 (2020) 18-44. https://doi.org/10.1016/j.jmst.2020.02.037

[21] Q. Shan, X. Mu, M. Alhabeb, C.E. Shuck, D. Pang, X. Zhao, X.F. Chu, Y. Wei, F. Du, G. Chen, Y. Gogotsi, Y. Gao, Y. Dall'Agnese, Two-dimensional vanadium carbide (V_2C) MXene as an electrode for supercapacitors with aqueous electrolytes, Electrochem. Commun. 96 (2018) 103-107. https://doi.org/10.1016/j.elecom.2018.10.012

[22] X. Wang, S. Lin, H. Tong, Y. Huang, P. Tong, B. Zhao, J. Dai, C. Liang, H. Wang, X. Zhu, Y. Sun, S. Dou, Two-dimensional V_4C_3 MXene as high-performance electrode materials for supercapacitors, Electrochim. Acta. 307 (2019) 414-421. https://doi.org/10.1016/j.electacta.2019.03.205

[23] S. Zhao, C. Chen, X. Zhao, X. Chu, F. Du, G. Chen, Y. Gogotsi, Y. Gao, Y. Dall'Agnese, Flexible $Nb_4C_3T_x$ film with large interlayer spacing for high-performance supercapacitors, Adv. Funct. Mater. 30 (2020) 1-8. https://doi.org/10.1002/adfm.202000815

[24] A.S. Etman, J. Halim, J. Rosen, Fabrication of $Mo_{1.33}CT_z$ (MXene)-cellulose freestanding electrodes for supercapacitor applications, Mater. Adv. 2 (2021) 743-753. https://doi.org/10.1039/D0MA00922A

[25] Y. Zhou, K. Maleski, B. Anasori, J.O. Thostenson, Y. Pang, Y. Feng, K. Zeng, C.B. Parker, S. Zauscher, Y. Gogotsi, J.T. Glass, C. Cao, $Ti_3C_2T_x$ MXene-reduced graphene oxide composite electrodes for stretchable supercapacitors, ACS Nano. 14 (2020) 3576-3586. https://doi.org/10.1021/acsnano.9b10066

[26] L. Qin, Q. Tao, A. El Ghazaly, J.f. Rodriguez, P.O.Å. Persson, J. Rosen, F. Zhang, High-performance ultrathin flexible solid-state supercapacitors based on solution processable Mo1.33C MXene and PEDOT:PSS, Adv. Funct. Mater. 28 (2018) 1-8. https://doi.org/10.1002/adfm.201703808

[27] J. Vyskočil, C.C.M. Martinez, K. Szőkölová, A. Dash, J.G. Julian, Z. Sofer, M. Pumera, 2D stacks of MXene Ti_3C_2 and 1T-Phase WS_2 with enhanced capacitive behavior, ChemElectroChem. 6 (2019) 3982-3986. https://doi.org/10.1002/celc.201900643

[28] X. Lu, J. Zhu, W. Wu, B. Zhang, Hierarchical architecture of $PANI@TiO_2/Ti_3C_2T_x$ ternary composite electrode for enhanced electrochemical performance, Electrochim. Acta. 228 (2017) 282-289. https://doi.org/10.1016/j.electacta.2017.01.025

[29] J. Zhu, X. Lu, L. Wang, Synthesis of a MoO_3/Ti_3C_2T: X composite with enhanced capacitive performance for supercapacitors, RSC Adv. 6 (2016) 98506-98513. https://doi.org/10.1039/C6RA15651G

[30] Y. Wang, H. Dou, J. Wang, B. Ding, Y. Xu, Z. Chang, X. Hao, Three-dimensional porous MXene/layered double hydroxide composite for high performance supercapacitors, J. Power Sources. 327 (2016) 221-228. https://doi.org/10.1016/j.jpowsour.2016.07.062

[31] N. Liu, Y. Gao, Recent progress in micro-supercapacitors with in-plane interdigital electrode architecture, Small. 13 (2017) 1-10. https://doi.org/10.1002/smll.201701989

[32] Z. Pan, F. Cao, X. Hu, X. Ji, A facile method for synthesizing CuS decorated Ti_3C_2 MXene with enhanced performance for asymmetric supercapacitors, J. Mater. Chem. A. 7 (2019) 8984-8992. https://doi.org/10.1039/C9TA00085B

[33] R. Ramachandran, K. Rajavel, W. Xuan, D. Lin, F. Wang, Influence of $Ti_3C_2T_x$ (MXene) intercalation pseudocapacitance on electrochemical performance of Co-MOF binder-free electrode, Ceram. Int. 44 (2018) 14425-14431. https://doi.org/10.1016/j.ceramint.2018.05.055

[34] R. Zou, H. Quan, M. Pan, S. Zhou, D. Chen, X. Luo, Self-assembled MXene $(Ti_3C_2T_x)/\alpha\text{-}Fe_2O_3$ nanocomposite as negative electrode material for supercapacitors, Electrochim. Acta. 292 (2018) 31-38. https://doi.org/10.1016/j.electacta.2018.09.149

[35] K. Zhao, H. Wang, C. Zhu, S. Lin, Z. Xu, X. Zhang, Free-standing MXene film modified by amorphous FeOOH quantum dots for high-performance asymmetric supercapacitor, Electrochim. Acta. 308 (2019) 1-8. https://doi.org/10.1016/j.electacta.2019.03.225

[36] M.Q. Zhao, C.E. Ren, Z. Ling, M.R. Lukatskaya, C. Zhang, K.L. Van Aken, M.W. Barsoum, Y. Gogotsi, Flexible MXene/carbon nanotube composite paper with high volumetric capacitance, Adv. Mater. 27 (2015) 339-345. https://doi.org/10.1002/adma.201404140

[37] M. Zhu, Y. Huang, Q. Deng, J. Zhou, Z. Pei, Q. Xue, Y. Huang, Z. Wang, H. Li, Q. Huang, C. Zhi, Highly flexible, freestanding supercapacitor electrode with enhanced performance obtained by hybridizing polypyrrole chains with MXene, Adv. Energy Mater. 6 (2016). https://doi.org/10.1002/aenm.201600969

[38] M. Boota, B. Anasori, C. Voigt, M.Q. Zhao, M.W. Barsoum, Y. Gogotsi, Pseudocapacitive electrodes produced by oxidant-free polymerization of pyrrole between the layers of 2d titanium carbide (MXene), Adv. Mater. 28 (2016) 1517-1522. https://doi.org/10.1002/adma.201504705

[39] Z. Ling, C.E. Ren, M.Q. Zhao, J. Yang, J.M. Giammarco, J. Qiu, M.W. Barsoum, Y. Gogotsi, Flexible and conductive MXene films and nanocomposites with high capacitance, Proc. Natl. Acad. Sci. U. S. A. 111 (2014) 16676-16681. https://doi.org/10.1073/pnas.1414215111

[40] Q. Wang, S. Wang, X. Guo, L. Ruan, N. Wei, Y. Ma, J. Li, M. Wang, W. Li, W. Zeng, MXene-reduced graphene oxide aerogel for aqueous zinc-ion hybrid supercapacitor with ultralong cycle life, Adv. Electron. Mater. 5 (2019) 1-8. https://doi.org/10.1002/aelm.201900537

[41] A. Vahidmohammadi, J. Moncada, H. Chen, E. Kayali, J. Orangi, C.A. Carrero, M. Beidaghi, Thick and freestanding MXene/PANI pseudocapacitive electrodes with ultrahigh specific capacitance, J. Mater. Chem. A. 6 (2018) 22123-22133. https://doi.org/10.1039/C8TA05807E

[42] N. Wang, J. Liu, Y. Zhao, M. Hu, R. Qin, G. Shan, Laser-cutting fabrication of Mxene-based flexible micro-supercapacitors with high areal capacitance, ChemNanoMat. 5 (2019) 658-665. https://doi.org/10.1002/cnma.201800674

[43] M. Ghidiu, M.R. Lukatskaya, M.Q. Zhao, Y. Gogotsi, M.W. Barsoum, Conductive two-dimensional titanium carbide "clay" with high volumetric capacitance, Nature. 516 (2015) 78-81. https://doi.org/10.1038/nature13970

[44] M. Hu, T. Hu, Z. Li, Y. Yang, R. Cheng, J. Yang, C. Cui, X. Wang, Surface functional groups and interlayer water determine the electrochemical capacitance of $Ti_3C_2T_x$ MXene, ACS Nano. 12 (2018) 3578-3586. https://doi.org/10.1021/acsnano.8b00676

[45] C.A. Voigt, M. Ghidiu, V. Natu, M.W. Barsoum, Anion Adsorption, $Ti_3C_2T_z$ MXene multilayers, and their effect on claylike swelling, J. Phys. Chem. C. 122 (2018) 23172-23179. https://doi.org/10.1021/acs.jpcc.8b07447

[46] S. Kajiyama, L. Szabova, H. Iinuma, A. Sugahara, K. Gotoh, K. Sodeyama, Y. Tateyama, M. Okubo, A. Yamada, Enhanced Li-Ion accessibility in MXene titanium carbide by steric chloride termination, Adv. Energy Mater. 7 (2017) 1-8. https://doi.org/10.1002/aenm.201601873

[47] J. Tang, T. Mathis, X. Zhong, X. Xiao, H. Wang, M. Anayee, F. Pan, B. Xu, Y. Gogotsi, Optimizing ion pathway in titanium carbide mxene for practical high-rate supercapacitor, Adv. Energy Mater. 11 (2021) 1-8. https://doi.org/10.1002/aenm.202003025

[48] M. Hu, Z. Li, G. Li, T. Hu, C. Zhang, X. Wang, All-solid-state flexible fiber-based MXene supercapacitors, Adv. Mater. Technol. 2 (2017) 1-6. https://doi.org/10.1002/admt.201700143

[49] W. Shao, M. Tebyetekerwa, I. Marriam, W. Li, Y. Wu, S. Peng, S. Ramakrishna, S. Yang, M. Zhu, Polyester@MXene nanofibers-based yarn electrodes, J. Power Sources. 396 (2018) 683-690. https://doi.org/10.1016/j.jpowsour.2018.06.084

[50] K. Krishnamoorthy, P. Pazhamalai, S. Sahoo, S.J. Kim, Titanium carbide sheet based high performance wire type solid state supercapacitors, J. Mater. Chem. A. 5 (2017) 5726-5736. https://doi.org/10.1039/C6TA11198J

[51] E.A. Mayerberger, O. Urbanek, R.M. McDaniel, R.M. Street, M.W. Barsoum, C.L. Schauer, Preparation and characterization of polymer-$Ti_3C_2T_x$ (MXene) composite nanofibers produced via electrospinning, J. Appl. Polym. Sci. 134 (2017) 1-7. https://doi.org/10.1002/app.45295

[52] A. Levitt, J. Zhang, G. Dion, Y. Gogotsi, J.M. Razal, MXene-based fibers, yarns, and fabrics for wearable energy storage devices, Adv. Funct. Mater. 30 (2020) 1-22. https://doi.org/10.1002/adfm.202000739

[53] H. Huang, H. Su, H. Zhang, L. Xu, X. Chu, C. Hu, H. Liu, N. Chen, F. Liu, W. Deng, B. Gu, H. Zhang, W. Yang, Extraordinary areal and volumetric performance of flexible solid-state micro-supercapacitors based on highly conductive freestanding $Ti_3C_2T_x$ films, Adv. Electron. Mater. 4 (2018) 1-9. https://doi.org/10.1002/aelm.201800179

[54] J. Zhou, J. Yu, L. Shi, Z. Wang, H. Liu, B. Yang, C. Li, C. Zhu, J. Xu, A Conductive and highly deformable all-pseudocapacitive composite paper as supercapacitor electrode with improved areal and volumetric capacitance, Small. 14 (2018) 1-9. https://doi.org/10.1002/smll.201803786

[55] N. Kurra, B. Ah qmed, Y. Gogotsi, H.N. Alshareef, MXene-on-Paper Coplanar Microsupercapacitors, Adv. Energy Mater. 6 (2016) 1-8. https://doi.org/10.1002/aenm.201601372

[56] P. Forouzandeh, S.C. Pillai, MXenes-based nanocomposites for supercapacitor applications, Curr. Opin. Chem. Eng. 33 (2021) 100710. https://doi.org/10.1016/j.coche.2021.100710

[57] Y. Li, Z. Lu, B. Xin, Y. Liu, Y. Cui, Y. Hu, All-solid-state flexible supercapacitor of carbonized MXene/cotton fabric for wearable energy storage, Appl. Surf. Sci. 528 (2020) 146975. https://doi.org/10.1016/j.apsusc.2020.146975

[58] T.A. Le, N.Q. Tran, Y. Hong, H. Lee, Intertwined titanium carbide MXene within a 3 D tangled polypyrrole nanowires matrix for enhanced supercapacitor performances, Chem. - A Eur. J. 25 (2019) 1037-1043. https://doi.org/10.1002/chem.201804291

[59] L. Sun, G. Song, Y. Sun, Q. Fu, C. Pan, MXene/N-doped carbon foam with three-dimensional hollow neuron-like architecture for freestanding, highly compressible all solid-state supercapacitors, ACS Appl. Mater. Interfaces. 12 (2020) 44777-44788. https://doi.org/10.1021/acsami.0c13059

[60] Y. Yue, N. Liu, Y. Ma, S. Wang, W. Liu, C. Luo, H. Zhang, F. Cheng, J. Rao, X. Hu, J. Su, Y. Gao, Highly self-healable 3D microsupercapacitor with MXene-graphene composite aerogel, ACS Nano. 12 (2018) 4224-4232. https://doi.org/10.1021/acsnano.7b07528

[61] N. Choudhary, C. Li, J. Moore, N. Nagaiah, L. Zhai, Y. Jung, J. Thomas, Asymmetric supercapacitor electrodes and devices, Adv. Mater. 29 (2017). https://doi.org/10.1002/adma.201605336

[62] K. Ghosh, M. Pumera, MXene and MoS_{3-x} Coated 3D-Printed Hybrid Electrode for Solid-State Asymmetric Supercapacitor, Small Methods. 5 (2021) 1-15. https://doi.org/10.1002/smtd.202100451

[63] M. Boota, M. Rajesh, M. Bécuwe, Multi-electron redox asymmetric supercapacitors based on quinone-coupled viologen derivatives and $Ti_3C_2T_x$ MXene, Mater. Today Energy. 18 (2020) 100532. https://doi.org/10.1016/j.mtener.2020.100532

[64] C. Couly, M. Alhabeb, K.L. Van Aken, N. Kurra, L. Gomes, A.M.N. Suárez, B. Anasori, H.N. Alshareef, Y. Gogotsi, Asymmetric flexible MXene-reduced graphene oxide micro-supercapacitor, Adv. Electron. Mater. 4 (2018) 1-8. https://doi.org/10.1002/aelm.201700339

[65] Q. Jiang, N. Kurra, M. Alhabeb, Y. Gogotsi, H.N. Alshareef, All pseudocapacitive MXene-RuO_2 asymmetric supercapacitors, Adv. Energy Mater. 8 (2018) 1-10. https://doi.org/10.1002/aenm.201703043

[66] A. El Ghazaly, W. Zheng, J. Halim, E.N. Tseng, P.O. Persson, B. Ahmed, J. Rosen, Enhanced supercapacitive performance of $Mo_{1.33}C$ MXene based asymmetric supercapacitors in lithium chloride electrolyte, Energy Storage Mater. 41 (2021) 203-208. https://doi.org/10.1016/j.ensm.2021.05.006

[67] J. Fu, L. Li, J.M. Yun, D. Lee, B.K. Ryu, K.H. Kim, Two-dimensional titanium carbide (MXene)-wrapped sisal-Like $NiCo_2S_4$ as positive electrode for high-performance hybrid pouch-type asymmetric supercapacitor, Chem. Eng. J. 375 (2019) 121939. https://doi.org/10.1016/j.cej.2019.121939

[68] J. Fu, J. Yun, S. Wu, L. Li, L. Yu, K.H. Kim, Architecturally robust graphene-encapsulated MXene Ti_2CT_x@polyaniline composite for high-performance pouch-type asymmetric supercapacitor, ACS Appl. Mater. Interfaces. 10 (2018) 34212-34221. https://doi.org/10.1021/acsami.8b10195

[69] Y.Y. Peng, B. Akuzum, N. Kurra, M.Q. Zhao, M. Alhabeb, B. Anasori, E.C. Kumbur, H.N. Alshareef, M.D. Ger, Y. Gogotsi, All-MXene (2D titanium carbide) solid-state microsupercapacitors for on-chip energy storage, Energy Environ. Sci. 9 (2016) 2847-2854. https://doi.org/10.1039/C6EE01717G

[70] Q. Li, Q. Wang, L. Li, L. Yang, Y. Wang, X. Wang, H.T. Fang, Femtosecond laser-etched MXene microsupercapacitors with double-side configuration via arbitrary on- and through-substrate connections, Adv. Energy Mater. 10 (2020) 1-8. https://doi.org/10.1002/aenm.202000470

[71] H. Li, X. Li, J. Liang, Y. Chen, Hydrous RuO_2 -decorated MXene coordinating with silver nanowire inks enabling fully printed micro-supercapacitors with extraordinary volumetric performance, Adv. Energy Mater. 9 (2019) 1-13. https://doi.org/10.1002/aenm.201803987

[72] J. Halim, M.R. Lukatskaya, K.M. Cook, J. Lu, C.R. Smith, L.Å. Näslund, S.J. May, L. Hultman, Y. Gogotsi, P. Eklund, M.W. Barsoum, Transparent conductive two-dimensional titanium carbide epitaxial thin films, Chem. Mater. 26 (2014) 2374-2381. https://doi.org/10.1021/cm500641a

[73] Y. Wen, T.E. Rufford, X. Chen, N. Li, M. Lyu, L. Dai, L. Wang, Nitrogen-doped $Ti_3C_2T_x$ MXene electrodes for high-performance supercapacitors, Nano Energy 38 (2017) 368-376. https://doi.org/10.1016/j.nanoen.2017.06.009

[74] C.J. Zhang, B. Anasori, A.S. -Ascaso, S.H. Park, N. McEvoy, A. Shmeliov, G.S. Duesberg, J.N. Coleman, Y. Gogotsi, V. Nicolosi, Transparent, flexible, and conductive 2D titanium carbide (MXene) films with high volumetric capacitance, Adv. Mater. 29 (2017) 1-9. https://doi.org/10.1002/adma.201702678

[75] K. Hantanasirisakul, M.Q. Zhao, P. Urbankowski, J. Halim, B. Anasori, S. Kota, C.E. Ren, M.W. Barsoum, Y. Gogotsi, Fabrication of $Ti_3C_2T_x$ MXene transparent thin films with tunable optoelectronic properties, Adv. Electron. Mater. 2 (2016) 1-7. https://doi.org/10.1002/aelm.201600050

[76] D. Wen, X. Wang, L. Liu, C. Hu, C. Sun, Y. Wu, Y. Zhao, J. Zhang, X. Liu, G. Ying, Inkjet printing transparent and conductive MXene ($Ti_3C_2T_x$) films: A strategy for flexible energy storage devices, ACS Appl. Mater. Interfaces 13 (2021) 17766-17780. https://doi.org/10.1021/acsami.1c00724

Recent Advances and Allied Applications of MXenes

Materials Research Foundations 155 (2024) 28-47

Materials Research Forum LLC

https://doi.org/10.21741/9781644902875-2

Chapter 2

Applications of MXenes in EMI shielding

Jhilmil Swapnalin[1], Bhargavi Koneru[1], Ramyakrishna Pothu[2], Ramachandra Naik[3],
Rajender Boddula[4], Ahmed Bahgat Radwan[4], Noora Al-Qahtani[4] and Prasun Banerjee[1]*

[1]Multiferroic and Magnetic Material Research Laboratory, Gandhi Institute of Technology and Management (GITAM) University, Bengaluru, Karnataka, India

[2]School of Physics and Electronics, College of Chemistry and Chemical Engineering, Hunan University, Changsha 410082, China

[3]Department of Physics, New Horizon College of Engineering, Bangalore 560103, India

[4]Center for Advanced Materials (CAM), Qatar University Doha 2713, Qatar

* pbanerje@gitam.edu

Abstract

The tremendous growth of wireless communication and smart nano-micro electronic devices has triggered electromagnetic interference (EMI) pollution in the environment. EMI is a significant problem concerning both animals and humankind. Various problems like data leakage, interferences from electronic devices and health hazards can be addressed with an efficient electromagnetic (EM) wave shielding material. In this chapter, a new research interest, 2D material, is studied to investigate their EMI shielding capacity. 2D MXene are transition metal carbides or nitrides obtained from the parent MAX phase resulting in orderly stacked layers. Better EMI shielding mechanism within 2D layers and enhanced absorption in various composites of MXene are summarized.

Keywords

MXene, 2D Materials, Ti3C2Tx, Intercalation, EMI Shielding

Contents

1. Introduction

In the recent era of scientific advancement in electric information technology like satellite communication, high-speed processors, and broadband networks, electromagnetic radiation is affecting adversely [1]. Electromagnetic pollution arising due to electromagnetic interference (EMI) is a major problem to tackle. The prime source of electromagnetic waves can be held to the man-made electronic gadgets used in the gigahertz (GHz) band range (Fig 1). According to previous reports, the electric and magnetic fields devised manually hamper the natural biological processes [2]. It is evident from the studies that the birds fail to tune their magnetic compass due to disruptive noise coming from EM waves [3]. Not only birds, but even electromagnetic radiation is also equally hazardous to mankind, causing DNA mutation to severe chronic illness [4,5]. The possible explanation of the radiation hazard to the body can be simplified from the physics point of view as the body tissue absorbs and scatters the electromagnetic wave. When the radiation penetrates the molecular tissues, the EM fields of the wave get converted to mechanical force. The water molecules and different regular ions present in the body get affected due to the field. The ions get reoriented, and the functionality of chemicals gets distorted. The overall response to radiation complicates the normal functioning of the health system [6].

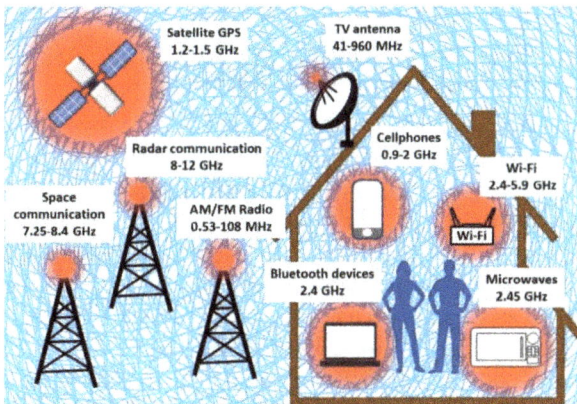

Figure 1. Representation of the EMI pollution surrounding mankind [7].

To combat this EMI pollution, a material with efficient EM attenuation being eco-friendly, cheap, lightweight, electrically and thermally conductive, and with high EMI shielding effectiveness (SE) is essential [7]. An EMI shielding material must also be a smart microwave absorber at high-temperature conditions. Moreover, to have super absorbing electromagnetic waves capacity, the material is demanded to have high magnetic and dielectric loss [8–12] thus, the electromagnetic impedance matching between complex permeability and permittivity should be optimum [13]. Various materials have been studied in this aspect owing to higher conductivity and EMI SE like r-GO, graphene nanosheets and graphene nanohybrids [14,15], composites with MWCNT [16], different conducting polymers [17–19], ceramics [20–22] and many magnetic materials like magnetic oxides, ferrites, metals, and ferroelectrics were studied to know their electromagnetic attenuation potential [23–27]. However, as per commercial demand, materials with lightweight, a thickness of nearly 1 mm, and EMI SE above 30 dB are preferable. In this aspect, very thin MXene films stand as a better alternative as per the literature survey.

The hydrophilicity and high metallic conductivity make MXenes suitable for EMI applications. As reported, shielding is subject to the abundance of free electrons available in materials that come in contact with penetrating radiation [28]. By increasing the conductivity of the material, we can tune the reflection from the material and interference scattering [29]. The floating functional groups on the layers of MXene make it eligible to provide better interaction with the incoming radiation. Thus, malleable, flexible, durable, and lightweight 1D, 2D, and 3D MXenes are the materials of interest and alternative to graphene as a better EMI- shielding option. MXenes, which show potential in electromagnetic wave shielding, have a broad scope of research and application in aerospace portable smart electronics [30–32].

2.   Electromagnetic interference shielding mechanism

EMI shielding property is basically the efficiency of any material to block or diminish the intensity of the propagating EM radiation. An electromagnetic wave is a plane wave propagating normally to the plane consisting of EM fields. The mechanism of EM screening is best understood by the plane-wave EM transmission theory given by Schelkunoff. When an EM wave penetrates through shielding material, then three types of phenomena occur, shield-effective reflection (SER), shield-effective multiple internal reflections (SEMR), and shield-effective absorption (SEA). As the EM wave penetrates through the material fraction of the wave gets reflected, another fraction gets absorbed and repeatedly reflected inside the shield material, and finally, the filtered wave passes through the shielding body called shield effective transmittance (SET), as shown in (Fig 2). Using these three components, the EMI SE is calculated. EMI SE expressed in dB determines the quality of the material for protection, where the absorption depends on the material type.

The first step of reflection occurs in a shielding material as the EM wave touches the external surface and interacts with active charge carriers. So, surfaces with active electrons and holes are highly conductive and weaken the incident EM wave by reflecting. The second step is the absorption of the projected wave. When the wave penetrates, the shield encounters both electric and magnetic dipoles, which extract wave energy for polarization [33–35] and diminish the wave strength. The third step is multiple internal reflections, a mechanism possible in layered materials. The waves get repeatedly reflected and absorbed within the layers. The resultant weak EM wave comes out as a transmitted wave.

The Electromagnetic SE of a material is determined by dividing the strength of incident EM waves without protecting the material by the intensity of the shielding material. EMI shielding can be done in two ways, namely near-field shielding when the source and shield are at a distance smaller than $\lambda/2\pi$ and at a larger distance, called far-field shielding. Different theories work in accordance with the field; thus, all parameters should be adequately taken to design a particular shielding composite.

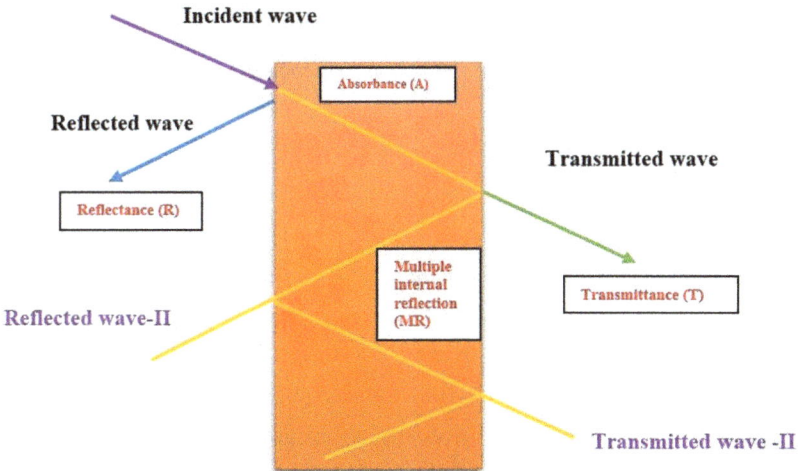

Figure 2. Schematic representation of phenomena occurring in EMI shielding.

The Schelkunoff formula for electromagnetic shielding materials forming a stable system is given as:

SET = SEMR +SER + SEA

Where SEA, SER, and SEMR are loss efficiency due to absorption, reflection, and multiple reflections, respectively and SET is the total output.

The following formula can do the calculation:

SER = 10lg σ_r /f μ_r + 168.2

SEA = 131.43t $\sqrt{}$ f $\sigma_r\mu_r$

and SEMR = 10lg [1 + $10^{-0.2SEA}$ − 2 × $10^{-0.1SEA}$ cos (0.23SEA)]

Where t is the shielding material thickness, and σ_r is the conductivity of the shielding material. SEMR is usually ignored when SET > 15 dB [36].

EMI SE in dB was also calculated by taking the incident and transmitted power of the wave as,

SET= -10 log (P_{out}/P_{in}) = -10 logT

SER= -10 log (1-R)

And in terms of incident and transmitted electric and magnetic fields intensities, these are given:

SET = 20 log10$|E_T|/|E_I|$

SET = 20log10$|H_T|/|H_I|$

Thus, SET= SER+SEA

Here T, R, A transmission coefficient, implies reflection, and absorption given as:

T+ R+A= 1

And A= (1-T-R/1-R) [37,38]

The transmitted electromagnetic radiation through a non-magnetic and conductive shielding material can be expressed using the attenuation rule with the Fresnel formula [39].

SET=10 log (1/T) =10 log $(E_i/E_t)^2$

= 20 log [{$(1+N)^2/4N$} *e^{-kd} {1-$(1-N/1+N)^2$ e^{2ikd}}]

N and d signify the shield's complex refractive index and thickness, respectively, while E_t and E_i denote the transmitted and incident fields, respectively. The imaginary portion of the refractive index is denoted by K.

Absorption loss arises due to ohmic loss in the material due to induced current and is independent of the source field. Furthermore, this loss of efficiency due to absorption and reflectance increases when the conductivity of the shielding material is increased, which in turn enhances the EMI shielding capacity.

Reflection and absorption losses can also be expressed as:

SER = $-10 \log_{10} (\sigma_T/16\omega\varepsilon\mu)$ and

SEA= $-8.686t (\sqrt{\sigma_T\omega\mu})$

Further,

SEA= $-8.686 (t/\delta)$

where, ω is the frequency, ε is permittivity, t is shielding thickness, δ is skin depth, and μ permeability.

When permeability and frequency are increased, the absorption loss increases. Also, with an increase in μ SER decreases, these factors constructively weaken the EM

wave penetration through the material by making it more absorptive and less reflective. A minimum thickness of shield is required, which can be characterized with respect to skin depth (δ). Once the EM passes through the material and when the skin depth is greater than the thickness, the wave's amplitude diminishes exponentially. As defined in the literature, the skin depth is the distance where the wave intensity is reduced by 1/e as compared to an initial value, and SEMR is the vital phenomenon. Whereas SER dominates when the material contains more free electrons like in metals, and SEA shielding is when materials show electric and magnetic dipole interactions [40,41].

3.   MXene for EMI shielding

Due to the remarkable features of MXene, a member of the two-dimension family is a hopeful material for EMI shielding material for future generation electronics technology. Titanium carbides and nitrides are known to be a better alternative for effective shielding due to their better conductivity, strength, flexibility, and easy solubility [42] MAX (Ti_3AlC_2) phase powder is carefully etched to bring the aluminum layer out and form MXene $Ti_3C_2T_x$. Here, T_x presents functional groups like O, OH, and F of the 2D layer. Another 2D material, graphene, with high carrier mobility, offers high electrical conductivity but is still not suitable to perform EMI shielding as compared to MXene [43].

Materials Research Forum LLC
https://doi.org/10.21741/9781644902875-2

In general, transverse electric mode waves travel parallel to the thin layered MXene plane, interfering due to multiple reflections impossible as the EM wave pathway is less than its wavelength. But the nanometre thickness of MXene sheets allows parallel long-range EM waves interaction with the MXene, due to its larger surface area. Thus, EM waves propagating perpendicular to the nanometre-thick MXene sheets exhibit the general phenomena of absorption and multiple reflections [44]. Furthermore, the surface decorated functional groups on MXene sheets are directed in the line of the applying electric field to get polarized and attenuate the incoming EM waves via polarization loss providing much absorption. And the electron-hopping mechanism occurring between the MXene layers also adds up to the attenuation process. Further, claimed by researchers, the spacing between the lamellar MXene layers may function as capacitors and attenuate the EM wave energy via capacitive dissipation [45,46].

The general mechanism of EMI shielding in the lamellar MXene layers was explained as the incident electromagnetic wave when interacting with the highly conductive MXene layers. Light is reflected from the surface first because of the abundance of free electrons. On the surface, local dipoles between Ti and other functional groups during EM wave interaction cause polarization losses. The filtered waves penetrate the layer, interact with more electrons, and conduct current, leading to ohmic loss. This process diminishes the energetic incident EM wave. Now, as the wave encounters another layer, which acts as a reflecting surface and attenuation repeats, these repeated layers in the material produce multiple internal reflections, and finally, the remnant waves get absorbed in the sheets. A very minute portion gets transmitted finally out of the sheets. Thus, in this way, MXene acts as an effective multi-layered EMI shielding material that locks the EM waves inside its layers, as shown in (Fig 3). Proving conducting titanium carbide, which offers good electronic coupling, the potential of being ultrathin and efficient EMI shielding material [47].

Recent Advances and Allied Applications of MXenes   Materials Research Forum LLC
Materials Research Foundations 155 (2024) 28-47   https://doi.org/10.21741/9781644902875-2

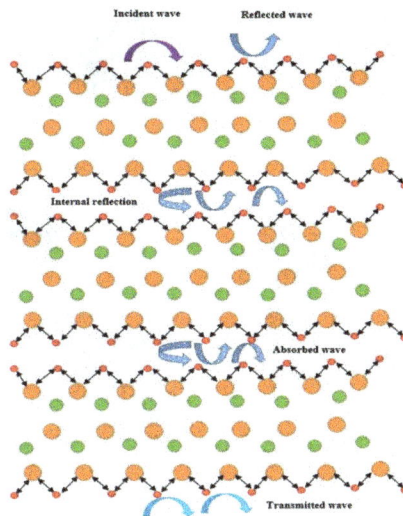

Figure 3. Schematic of EMI shielding mechanism occurring in MXene.

A 3D conducting network was made via a hydrothermal reaction between $Ti_3C_2T_x$ and graphene oxide (GO) with successive unidirectional freezing and freeze-drying. The aerogel hybrid of reduced GO formed the core, and the MXene sheets served as the outer shell. The electrical conductivity (696 Sm^{-1}) with an EMI SE 56.4 dB, showed better results as compared to previously reported polymer/MXene nanocomposite [48]. Nanostructured MXene combined with carbon nanotubes and polymers using a spin spray layer-by-layer process (SSLBL) offers a layered composite. The semi-transparent layer-by-layer MXene-MWCNT composite provided controlled cross-functionalities like transparency, conductivity, and thickness with high flexibility and stability. Greater EMI shielding productiveness of 58187.0 $dB.cm^2/g$ was reported, where the absorption phenomenon is the lead in the shielding mechanism [49]. Polydimethylsiloxane (PDMS)- coated MXene foams prepared using aerogel of 3D $Ti_3C_2T_x$ with sodium alginate (SA) resulted in flexible, compressible, and durable EMI shielding of 70.5 dB (Fig 4) [50].

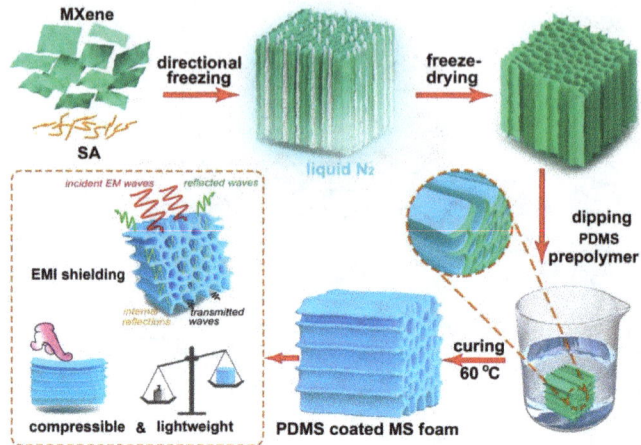

Figure 4. Representation of PDMS coated MXene/SA composite formation [50].

With the 2D metal carbides and nitrides, three-dimensional porous aerogel structures of a series of MXene were synthesized via bidirectional freeze-casting. At a definite thickness of 1 mm and extremely low density of 0.0055 gm, EMI SE of the Ti_2CT_x aerogel was enhanced to 48.5 dB, thus showing compressible MXene being a controlled method to tune EM wave absorption and reflection in the shielding material [51]. Researchers have reported the EMI shielding efficiency of monolayer MXene assemblies to be 20%. When 24 monolayers were stacked, making a thickness in the nanometre range, SE increased to 99%, which is around 20 dB owing to the metallic behaviour of 2D carbides [52].

3.1 Recent progress in EMI shielding performance of different MXenes composites

Z. Fan, et al., to check the poor EMI shielding capacity of 3D graphene fabricated by graphene oxide (GO), proposed a composite with MXene. Two-dimensional MXene nanosheets with excellent conductivity were introduced into GO to prepare an MXene/graphene hybrid foam. The resultant lightweight, conductive, porous structure of MX-rGO showed an EMI SE of 50.7 dB. It also displayed high specific shielding efficiency (SSE) of 6217 cm^3/g at lower thickness. Thus, the interconnected porous network offers a larger surface area, thus enabling rapid EM wave attenuation in the composite [53]. Hydrophilic MXene is very difficult to use in damp conditions. Thus, hydrophobic MXene-foam is proposed to be prepared by a hydrazine-propelled foaming procedure. The resultant composite exhibited elevated shelf-life with properly stacked hydrophobic layers. This composite turned out to be lightweight, flexy, and hydrophobic, with an improved EMI SE of nearly 70 dB [54].

Various MXene/PANI composites are prepared through various methods like in situ polymerization, vacuum-assisted filtration, spray coating, and hydrothermal synthesis, exhibiting excellent physical and chemical properties. Due to the characteristics like good conductivity and dielectric loss MXene/PANI shows excellent potential to be used in addressing the EM wave pollution issue. Constructing the firm 2D hierarchical layers and introducing a polymer PANI within them proved the composite a better candidate for EM wave absorption [55]. Scalable production of flexible, controllable, and durable MXene composite was reported by Y. Li et al. Mxene nanosheets with polyvinylidene fluoride (PVDF) were fabricated using the blade coating method. Composite attenuated EM wave by interfacial polarization and multiple reflections within the 2D network showing EMI SE (24.9–42.9 dB) and SSE/t of 19504.8 dB cm2 g−1 at only 17 μm thickness [56].

Another group of researchers to examine the EMI shielding performance of $Ti_3C_2T_x$ checked their susceptibility to oxidation. Due to oxidation in a humid environment, 2D layers get disassembled, weakening the mechanical support, functional groups effect, and electrical conductivity. MXene-based on imidazolium ionic liquid (IL) was proposed, providing good chemical and mechanical stability. Here, the surface chemistry of MXene sheets and the radical-scavenging ability of IL were explored to get high tensile MXene-IL film (75.9 ± 4.9 MPa). This IL-treated MXene was found to be stable in aqueous mediums, high temperatures, and rough environmental conditions, hence an efficient composite to reduce EMI pollution [57].

L. Wang et al. reported $Ti_3C_2T_x$/epoxy composite using the solution casting method. The annealed nanocomposite was analyzed through various characterizations like XRD, SEM, FTIR, TEM, and AFM, confirming 2D lamellar layer formation and reduced polar groups on the surface due to annealing. At 5%, the annealed composite exhibited great hardness with an optimal Young's modulus of 4.32 GPa. 15% annealed $Ti_3C_2T_x$/ epoxy shows better electrical conductivity of 105 S/m and EMI SE around 41 dB than $Ti_3C_2T_x$/ epoxy, indicating a good EMI shielding nanocomposite for application [58]. Polymeric nanocomposite like M-X@chitosan (CS) was fabricated in layered order, which offered excellent thermal conductivity and EMI SE of 40.8 dB at a thickness of 35 mm [59]. In another study, pristine MXene was phosphorylated and deposited on polypropylene (PP), and MX@PP nanocomposite was produced via compression molding. As a result, a durable, oxidation resistant with enhanced conductivity of around 36,700 S/m and EMI SE of nearly 90 dB was recorded [60].

MXene with aerogels is a new type of composite evaluated for EMI shielding efficiency. MXene infused with super-flat cellulose nanofibrils (CNFs) results in a less dense, flexible, and strong aerogel. Here, the 1D CNFs with 2D MXene (2 mm thick) offer a high EMI SE of around 35.5 dB as compared to the 2 mm pristine MXene aerogel (25.5 dB) [61]. In another report, MXene was amalgamated with phosphated lignocellulose nanofibrils (PLCNFs) and gelatin to form porous MX@P@G aerogel. This aerogel composite exhibited specific shielding efficiency of ($14,230$ dB cm^2 g^{-1}) with very less density [62].

Also, 2D MXene with 1D silver nanowires together formed 3D MX@AgNWs aerogels. The fabricated aerogels are mixed with a polymer like Epoxy to get the final product as MX@AgNWs@Epoxy EMI shielding nanocomposites. It showed elevated electrical conductivity of 1532 S/m and EMI SE value (94.1 dB) at 3 mm thickness [63]. Another nanoarchitecture film with efficient EMI shielding performance was reported using the casting method. Ternary MX@GN@Fe3O4@PVA hybrid was found flexy and well compatible with the PVA matrix giving EMI SE of 36 dB at 1 mm thickness [64].

Furthermore, many research studies are being carried out on optimizing the EMI shielding efficiency using 2D material MXene. Various materials like intrinsic conductive polymers (ICP), cement, textiles, carbon black, and nonconducting polymer materials are used to enhance EMI SE with $Ti_3C_2T_x$. Better substitutes, fillers, and additives to form MXene hybrid opens a new potential research benefit for the aerospace industry. The last 5 years' progress on EMI shielding and electrical conductivities of MXenes is shown in Table 1.

Table 1. Progress on EMI shielding and electrical conductivities on MXenes in the last 5 years.

Sl.No.	Name of composite	Year	Electrical conductivity [s/m]	EMI SE [dB]	References
a.	Hydrophobic MXene foam	2017	58000	70	[65]
b.	$Ti_3C_2T_x$-carbon fabric	2018	850	43.2	[66]
c.	$Ti_3C_2T_x$/TOCNF	2019	2837	39.6	[67]
d.	Poly (vinyl alcohol)/MXene	2020	716	44.4	[68]
e.	Bilayer MXene/cellulose paper sheets	2021	5935.4	60.1	[69]

Conclusion

A comprehensive study of the preparation of different types of MXene from the MAX phase is done to study the EMI screening capacity of MXene and its composites. After investigating various MXene forms, it is concluded that the EMI SE of two-dimensional materials is high as per the recent research on better EM wave absorbing material. Pristine $Ti_3C_2T_x$ and their different composites showed great potential for better EMA material providing a solution to mitigate EMI pollution. The metallic nature, high Conductivity, and surface-decorated functional groups of MXene, when incorporated with suitable material, enhanced the attenuation of incident EM waves. EMI shielding mechanism was executed well and was two-fold in the accordion-type material. The 2D lamellar structure offered polarization loss on the surface, ohmic loss once the EM wave penetrates the layer, and finally, better absorption within the subsequent layers. The transmitted wave was reported to be very weak, providing nearly total shielding of the electromagnetic waves. Subsequent experiments were carried out to fabricate a better absorbing material using 2D metal carbides and nitrides by increasing M or X in the composition of the MAX phase. A step forward was taken to introduce a 3D MXene porous structure to decrease secondary transmission pollution, further optimize the absorption and reduce reflection. Various matrix materials like sponges, foams, and aerogels have been incorporated into the layers for better results. When magnetic materials were encapsulated in the 2D layers, due to interface polarization and magnetic loss offered by the composite, they increased the absorption tendency of MXene many folds. Thus, recent research suggests magnetic materials with MXene make better EM wave shielding composite in contrast to other forms. An advantage of MXene composite was attributed to the extraction of properties of both materials leading to higher attenuation by a mechanically and thermally stable shield. Hence, 2D MXene is an intriguing topic due to the rapid growth in information and communication technology, the requirement of energy conversion, upcoming 5G networking, wearable, lightweight-efficient devices, and ultimately providing a green electromagnetic environment.

Acknowledgments

P. Banerjee thanks SERB, India, for a TAR/2021/000032 research grant. J Swapnalin and B Koneru thank GITAM University for a T. A. and SJSGC (F. No. 82-7/2022(SA-III)) UGC JRF research fellowship.

References

[1] M.M. Lu, M.S. Cao, Y.H. Chen, W.Q. Cao, J. Liu, H.L. Shi, D.Q. Zhang, W.Z. Wang, J. Yuan, Multiscale assembly of grape-like ferroferric oxide and carbon nanotubes: A smart absorber prototype varying temperature to tune intensities, ACS Appl. Mater. Interfaces. 7 (2015) 19408-19415. https://doi.org/10.1021/acsami.5b05595

[2] N.S. Kumar, R.P. Suvarna, K.C.B. Naidu, P. Banerjee, A. Ratnamala, H. Manjunatha, A review on biological and biomimetic materials and their applications, Appl. Phys. A. 126 (2020) 1-18. https://doi.org/10.1007/s00339-019-3176-6

[3] S. Engels, N.-L. Schneider, N. Lefeldt, C.M. Hein, M. Zapka, A. Michalik, D. Elbers, A. Kittel, P.J. Hore, H. Mouritsen, Anthropogenic electromagnetic noise disrupts magnetic compass orientation in a migratory bird, Nature. 509 (2014) 353-356. https://doi.org/10.1038/nature13290

[4] I.S. Group, Brain tumour risk in relation to mobile telephone use: Results of the INTERPHONE international case-control study, Int. J. Epidemiol. 39 (2010) 675-694. https://doi.org/10.1093/ije/dyq079

[5] M. Elwood, A.W. Wood, Health effects of radiofrequency electromagnetic energy, Health N Hav. 132 (2019).

[6] R.S. Kasevich, Cellphones, radars, and health [Speakout], IEEE Spectr. 39 (2002) 15-16. https://doi.org/10.1109/MSPEC.2002.1021945

[7] S. Gupta, N.-H. Tai, Carbon materials and their composites for electromagnetic interference shielding effectiveness in X-band, Carbon N. Y. 152 (2019) 159-187. https://doi.org/10.1016/j.carbon.2019.06.002

[8] P. Banerjee, A.F. Junior, D.B. Basha, K.C. Naidu, Magnetic nanomaterials for spintronics, Magnetochem: Mater. Appl. 66 (2020) 323.

[9] M. Prakash, N.S. Kumar, K.C.B. Naidu, M. Sarma, P. Banerjee, R.J. Kumar, R. Pothu, R. Boddula, Electrode materials for K-ion batteries and applications, in: Inamuddin, R. Boddula, A.M. Asiri (Eds.), Potassium-Ion Batteries: Materials and applications, 2020, pp.123-136. https://doi.org/10.1002/9781119663287.ch5

[10] L.G. Pereira, R.L. de S. Silva, P. Banerjee, A. Franco Jr, Role of Gd3+ ions on the magnetic hyperthermic behavior of anisotropic CoFe2O4 nanoparticles, Phys. B Condens. Matter. 587 (2020) 412140. https://doi.org/10.1016/j.physb.2020.412140

[11] S.O. de Lira, R.L. de S. Silva, P. Banerjee, A. Franco Jr, Effects of defect dipoles on the colossal permittivity of ambipolar co-doped rutile TiO_2 ceramics, J. Phys. Chem. Solids. 143 (2020) 109456. https://doi.org/10.1016/j.jpcs.2020.109456

[12] K. Srinivas, K.C.B. Naidu, G. Balakrishna, B.V.S. Reddy, N.S. Kumar, S. Ramesh, P. Banerjee, D.B. Basha, Magnetic nanomaterials for supercapacitors, Magnetochem. À Mater. Appl. 66 (2020) 259.

[13] G. Wang, L. Wang, L.H. Mark, V. Shaayegan, G. Wang, H. Li, G. Zhao, C.B. Park, Ultralow-threshold and lightweight biodegradable porous PLA/MWCNT with segregated conductive networks for high-performance thermal insulation and electromagnetic interference shielding applications, ACS Appl. Mater. Interfaces. 10 (2018) 1195-1203. https://doi.org/10.1021/acsami.7b14111

[14] G. Wang, X. Peng, L. Yu, G. Wan, S. Lin, Y. Qin, Enhanced microwave absorption of ZnO coated with Ni nanoparticles produced by atomic layer deposition, J. Mater. Chem. A Mater. 3 (2015) 2734-2740. https://doi.org/10.1039/C4TA06053A

[15] B. Wen, X.X. Wang, W.Q. Cao, H.L. Shi, M.M. Lu, G. Wang, H.B. Jin, W.Z. Wang, J. Yuan, M.S. Cao, Reduced graphene oxides: The thinnest and most lightweight materials with highly efficient microwave attenuation performances of the carbon world, Nanoscale. 6 (2014) 5754-5761. https://doi.org/10.1039/C3NR06717C

[16] W.-Q. Cao, X.-X. Wang, J. Yuan, W.-Z. Wang, M.-S. Cao, Temperature dependent microwave absorption of ultrathin graphene composites, J. Mater. Chem. C Mater. 3 (2015) 10017-10022. https://doi.org/10.1039/C5TC02185E

[17] V.K. Sachdev, S.K. Sharma, M. Tomar, V. Gupta, R.P. Tandon, EMI shielding of MWCNT/ABS nanocomposites in contrast to graphite/ABS composites and MWCNT/PS nanocomposites, RSC Adv. 6 (2016) 45049-45058. https://doi.org/10.1039/C6RA04200G

[18] K. Lakshmi, H. John, K.T. Mathew, R. Joseph, K.E. George, Microwave absorption, reflection and EMI shielding of PU-PANI composite, Acta. Mater. 57 (2009) 371-375. https://doi.org/10.1016/j.actamat.2008.09.018

[19] L. Lyu, J. Liu, H. Liu, C. Liu, Y. Lu, K. Sun, R. Fan, N. Wang, N. Lu, Z. Guo, An overview of electrically conductive polymer nanocomposites toward electromagnetic interference shielding, Eng. Sci. 2 (2018) 26-42.

[20] N.S. Kumar, K.C.B. Naidu, P. Banerjee, H. Manjunatha, A. Ratnamala, S. Janardan, Advanced ceramics for Microwave Absorber Applications, Appl. Adv. Ceram. Sci. Technol. Med. 3 (2020) 51. https://doi.org/10.2174/9789811478192120030007

[21] P. Banerjee, A. Franco, K.C.B. Naidu, Advanced ceramics for ferroelectric devices, Appl. Adv. Ceram. Sci. Technol. Med. 3 (2020) 95. https://doi.org/10.2174/9789811478192120030010

[22] G.G. Miranda, R.L. de S. Silva, P. Banerjee, A. Franco Jr, Role of Ga presence into the heterojunction of metal oxide semiconductor on the stability and tunability ZnO ceramics, Ceram Int. 46 (2020) 23390-23396. https://doi.org/10.1016/j.ceramint.2020.06.022

[23] M. Wang, X.-H. Tang, J.-H. Cai, H. Wu, J.-B. Shen, S.-Y. Guo, Construction, mechanism and prospective of conductive polymer composites with multiple interfaces for electromagnetic interference shielding: A review, Carbon N Y. 177 (2021) 377-402. https://doi.org/10.1016/j.carbon.2021.02.047

[24] Q. Liu, Q. Cao, X. Zhao, H. Bi, C. Wang, D.S. Wu, R. Che, Insights into size-dominant magnetic microwave absorption properties of CoNi microflowers via off-axis electron holography, ACS Appl. Mater. Interfaces. 7 (2015) 4233-4240. https://doi.org/10.1021/am508527s

[25] Z. Yu, Z. Yao, N. Zhang, Z. Wang, C. Li, X. Han, X. Wu, Z. Jiang, Electric field-induced synthesis of dendritic nanostructured α-Fe for electromagnetic absorption application, J. Mater. Chem. A Mater. 1 (2013) 4571-4576. https://doi.org/10.1039/c3ta01641b

[26] C. Wang, X. Han, X. Zhang, S. Hu, T. Zhang, J. Wang, Y. Du, X. Wang, P. Xu, Controlled synthesis and morphology-dependent electromagnetic properties of hierarchical cobalt assemblies, J. Phys. Chem. C. 114 (2010) 14826-14830. https://doi.org/10.1021/jp1050386

[27] J. Liu, R. Che, H. Chen, F. Zhang, F. Xia, Q. Wu, M. Wang, Microwave absorption enhancement of multifunctional composite microspheres with spinel Fe_3O_4 cores and anatase TiO_2 shells, Small. 8 (2012) 1214-1221. https://doi.org/10.1002/smll.201102245

[28] L. Liu, M. Flores, N. Newman, Microwave loss in the high-performance dielectric Ba (Zn1/3 Ta2/3) O3 at 4.2 K, Phys Rev Lett. 109 (2012) 257601.

[29] H. Wang, K. Teng, C. Chen, X. Li, Z. Xu, L. Chen, H. Fu, L. Kuang, M. Ma, L. Zhao, Conductivity and electromagnetic interference shielding of graphene-based architectures using MWCNTs as free radical scavenger in gamma-irradiation, Mater. Lett. 186 (2017) 78-81. https://doi.org/10.1016/j.matlet.2016.09.086

[30] P. Banerjee, A. Franco Jr, K.C.B. Naidu, D.B. Basha, R. Pothu, R. Boddula, Active Materials for Flexible K-Ion Batteries, in: Inamuddin, R. Boddula, A.M. Asiri (Eds.), Potassium-Ion Batteries: Materials and Applications, 2020, pp. 137-145. https://doi.org/10.1002/9781119663287.ch6

[31] P. Banerjee, A. Franco Jr, R. Boddula, K.C.B. Naidu, R. Pothu, Carbon Nanomaterials for Zn-Ion Batteries, in: Inamuddin, R. Boddula, A.M. Asiri (Eds.), Zinc Batteries: Basics, Developments, and Applications, 2020, pp. 1-9. https://doi.org/10.1002/9781119662433.ch1

[32] P. Banerjee, N.S. Kumar, K.C.B. Naidu, A. Franco, R. Dachepalli, Stability of 2D and 3D perovskites due to inhibition of light-induced decomposition, J. Electron. Mater. 49 (2020) 7072-7084. https://doi.org/10.1007/s11664-020-08435-w

[33] S.O. de Lira, R.L. de S. Silva, P. Banerjee, A. Franco Jr, Effects of defect dipoles on the colossal permittivity of ambipolar co-doped rutile TiO_2 ceramics, J. Phys. Chem. Solids. 143 (2020) 109456. https://doi.org/10.1016/j.jpcs.2020.109456

[34] P. Banerjee, N.S. Kumar, A. Franco Jr, A.K. Swain, K. Chandra Babu Naidu, Insights into the dielectric loss mechanism of bianisotropic FeSi/SiC composite materials, ACS Omega. 5 (2020) 25968-25972. https://doi.org/10.1021/acsomega.0c03409

[35] P. Banerjee, A.F. Junior, D.B. Basha, K.C. Naidu, Magnetic nanomaterials for spintronics, Magnetochem: Mater. Appl. 66 (2020) 323.

[36] H. Wang, S. Li, M. Liu, J. Li, X. Zhou, Review on shielding mechanism and structural design of electromagnetic interference shielding composites, Macromol. Mater. Eng. 306 (2021) 2100032. https://doi.org/10.1002/mame.202100032

[37] K. Nasouri, A.M. Shoushtari, M.R.M. Mojtahedi, Theoretical and experimental studies on EMI shielding mechanisms of multi-walled carbon nanotubes reinforced high performance composite nanofibers, J. Polym. Res. 23 (2016) 1-8. https://doi.org/10.1007/s10965-016-0943-3

[38] Y.-J. Wan, P.-L. Zhu, S.-H. Yu, R. Sun, C.-P. Wong, W.-H. Liao, Graphene paper for exceptional EMI shielding performance using large-sized graphene oxide sheets and doping strategy, Carbon N Y. 122 (2017) 74-81. https://doi.org/10.1016/j.carbon.2017.06.042

[39] Z. Li, Z. Wang, W. Lu, B. Hou, Theoretical study of electromagnetic interference shielding of 2D MXenes films, Metals (Basel). 8 (2018) 652. https://doi.org/10.3390/met8080652

[40] D.D.L. Chung, Materials for electromagnetic interference shielding, J Mater. Eng. Perform. 9 (2000) 350-354. https://doi.org/10.1361/105994900770346042

[41] A. Joshi, S. Datar, Carbon nanostructure composite for electromagnetic interference shielding, Pramana. 84 (2015) 1099-1116. https://doi.org/10.1007/s12043-015-1005-9

[42] X. Chen, Y. Zhao, L. Li, Y. Wang, J. Wang, J. Xiong, S. Du, P. Zhang, X. Shi, J. Yu, MXene/polymer nanocomposites: preparation, properties, and applications, Polym Rev. 61 (2021) 80-115. https://doi.org/10.1080/15583724.2020.1729179

[43] P. Kumar, F. Shahzad, S. Yu, S.M. Hong, Y.-H. Kim, C.M. Koo, Large-area reduced graphene oxide thin film with excellent thermal conductivity and electromagnetic interference shielding effectiveness, Carbon N Y. 94 (2015) 494-500. https://doi.org/10.1016/j.carbon.2015.07.032

[44] M. Born, E. Wolf, Principles of optics: Electromagnetic theory of propagation, interference and diffraction of light, Elsevier, 2013.

[45] B. Wen, M. Cao, M. Lu, W. Cao, H. Shi, J. Liu, X. Wang, H. Jin, X. Fang, W. Wang, Reduced graphene oxides: Light-weight and high-efficiency electromagnetic interference shielding at elevated temperatures, Adv. Mater. 26 (2014) 3484-3489. https://doi.org/10.1002/adma.201400108

[46] M. Han, X. Yin, H. Wu, Z. Hou, C. Song, X. Li, L. Zhang, L. Cheng, Ti3C2 MXenes with modified surface for high-performance electromagnetic absorption and shielding in the X-band, ACS Appl. Mater. Interfaces. 8 (2016) 21011-21019. https://doi.org/10.1021/acsami.6b06455

[47] F. Shahzad, M. Alhabeb, C.B. Hatter, B. Anasori, S. Man Hong, C.M. Koo, Y. Gogotsi, Electromagnetic interference shielding with 2D transition metal carbides (MXenes), Science 353 (2016) 1137-1140. https://doi.org/10.1126/science.aag2421

[48] S. Zhao, H.-B. Zhang, J.-Q. Luo, Q.-W. Wang, B. Xu, S. Hong, Z.-Z. Yu, Highly electrically conductive three-dimensional Ti3C2Tx MXene/reduced graphene oxide hybrid aerogels with excellent electromagnetic interference shielding performances, ACS Nano. 12 (2018) 11193-11202. https://doi.org/10.1021/acsnano.8b05739

[49] G. Weng, J. Li, M. Alhabeb, C. Karpovich, H. Wang, J. Lipton, K. Maleski, J. Kong, E. Shaulsky, M. Elimelech, Layer-by-layer assembly of cross-functional semi-transparent MXene-carbon nanotubes composite films for next-generation electromagnetic interference shielding, Adv. Funct. Mater. 28 (2018) 1803360. https://doi.org/10.1002/adfm.201803360

[50] X. Wu, B. Han, H.-B. Zhang, X. Xie, T. Tu, Y. Zhang, Y. Dai, R. Yang, Z.-Z. Yu, Compressible, durable and conductive polydimethylsiloxane-coated MXene foams for high-performance electromagnetic interference shielding, Chem. Eng. J. 381 (2020) 122622. https://doi.org/10.1016/j.cej.2019.122622

[51] M. Han, X. Yin, K. Hantanasirisakul, X. Li, A. Iqbal, C.B. Hatter, B. Anasori, C.M. Koo, T. Torita, Y. Soda, Anisotropic MXene aerogels with a mechanically tunable ratio of electromagnetic wave reflection to absorption, Adv. Opt. Mater. 7 (2019) 1900267. https://doi.org/10.1002/adom.201900267

[52] T. Yun, H. Kim, A. Iqbal, Y.S. Cho, G.S. Lee, M. Kim, S.J. Kim, D. Kim, Y. Gogotsi, S.O. Kim, Electromagnetic shielding of monolayer MXene assemblies, Adv. Mater. 32 (2020) 1906769. https://doi.org/10.1002/adma.201906769

[53] Z. Fan, D. Wang, Y. Yuan, Y. Wang, Z. Cheng, Y. Liu, Z. Xie, A lightweight and conductive MXene/graphene hybrid foam for superior electromagnetic interference shielding, Chem. Eng. J. 381 (2020) 122696. https://doi.org/10.1016/j.cej.2019.122696

[54] C.I. Idumah, Recent advancements in electromagnetic interference shielding of polymer and Mxene nanocomposites, Polym-Plastics Technol Mater. 62 (2023) 19-53. https://doi.org/10.1080/25740881.2022.2089581

[55] Z. He, H. Xie, H. Wu, J. Chen, S. Ma, X. Duan, A. Chen, Z. Kong, Recent advances in MXene/polyaniline-based composites for electrochemical devices and electromagnetic interference shielding applications, ACS Omega. 6 (2021) 22468-22477. https://doi.org/10.1021/acsomega.1c02996

[56] Y. Li, B. Zhou, Y. Shen, C. He, B. Wang, C. Liu, Y. Feng, C. Shen, Scalable manufacturing of flexible, durable Ti3C2Tx MXene/Polyvinylidene fluoride film for multifunctional electromagnetic interference shielding and electro/photo-thermal conversion applications, Compos B Eng. 217 (2021) 108902. https://doi.org/10.1016/j.compositesb.2021.108902

[57] Y.-J. Wan, K. Rajavel, X.-M. Li, X.-Y. Wang, S.-Y. Liao, Z.-Q. Lin, P.-L. Zhu, R. Sun, C.-P. Wong, Electromagnetic interference shielding of Ti3C2Tx MXene modified by ionic liquid for high chemical stability and excellent mechanical strength, Chem. Eng. J. 408 (2021) 127303. https://doi.org/10.1016/j.cej.2020.127303

[58] L. Wang, L. Chen, P. Song, C. Liang, Y. Lu, H. Qiu, Y. Zhang, J. Kong, J. Gu, Fabrication on the annealed Ti3C2Tx MXene/Epoxy nanocomposites for electromagnetic interference shielding application, Compos. B Eng. 171 (2019) 111-118. https://doi.org/10.1016/j.compositesb.2019.04.050

[59] Z. Tan, H. Zhao, F. Sun, L. Ran, L. Yi, L. Zhao, J. Wu, Fabrication of Chitosan/MXene multilayered film based on layer-by-layer assembly: Toward enhanced electromagnetic interference shielding and thermal management capacity, Compos. Part A Appl. Sci. Manuf. 155 (2022) 106809. https://doi.org/10.1016/j.compositesa.2022.106809

[60] T. Tang, S. Wang, Y. Jiang, Z. Xu, Y. Chen, T. Peng, F. Khan, J. Feng, P. Song, Y. Zhao, Flexible and flame-retarding phosphorylated MXene/polypropylene composites for efficient electromagnetic interference shielding, J. Mater. Sci. Technol. 111 (2022) 66-75. https://doi.org/10.1016/j.jmst.2021.08.091

[61] Z. Zeng, C. Wang, G. Siqueira, D. Han, A. Huch, S. Abdolhosseinzadeh, J. Heier, F. Nüesch, C. Zhang, G. Nyström, Nanocellulose-MXene biomimetic aerogels with orientation-tunable electromagnetic interference shielding performance, Adv. Sci. 7 (2020) 2000979. https://doi.org/10.1002/advs.202000979

[62] Y. Li, Y. Chen, X. He, Z. Xiang, T. Heinze, H. Qi, Lignocellulose nanofibril/gelatin/MXene composite aerogel with fire-warning properties for enhanced electromagnetic interference shielding performance, Chem. Eng. J. 431 (2022) 133907. https://doi.org/10.1016/j.cej.2021.133907

[63] H. Liu, Z. Huang, T. Chen, X. Su, Y. Liu, R. Fu, Construction of 3D MXene/Silver nanowires aerogel reinforced polymer composites for extraordinary electromagnetic interference shielding and thermal conductivity, Chem. Eng. J. 427 (2022) 131540. https://doi.org/10.1016/j.cej.2021.131540

[64] Y. Yao, S. Jin, M. Wang, F. Gao, B. Xu, X. Lv, Q. Shu, MXene hybrid polyvinyl alcohol flexible composite films for electromagnetic interference shielding, Appl. Surf. Sci. 578 (2022) 152007. https://doi.org/10.1016/j.apsusc.2021.152007

[65] J. Liu, H. Zhang, R. Sun, Y. Liu, Z. Liu, A. Zhou, Z. Yu, Hydrophobic, flexible, and lightweight MXene foams for high-performance electromagnetic-interference shielding, Adv. Mater. 29 (2017) 1702367. https://doi.org/10.1002/adma.201702367

[66] K. Raagulan, R. Braveenth, H.J. Jang, Y.S. Lee, C. Yang, B.M. Kim, J.J. Moon, K.Y. Chai, Fabrication of nonwetting flexible free-standing MXene-carbon fabric for electromagnetic shielding in S-band region, Bull Korean Chem. Soc. 39 (2018) 1412-1419. https://doi.org/10.1002/bkcs.11616

[67] Z. Zhan, Q. Song, Z. Zhou, C. Lu, Ultrastrong and conductive MXene/cellulose nanofiber films enhanced by hierarchical nano-architecture and interfacial interaction for flexible electromagnetic interference shielding, J. Mater. Chem. C Mater. 7 (2019) 9820-9829. https://doi.org/10.1039/C9TC03309B

[68] X. Jin, J. Wang, L. Dai, X. Liu, L. Li, Y. Yang, Y. Cao, W. Wang, H. Wu, S. Guo, Flame-retardant poly (vinyl alcohol)/MXene multilayered films with outstanding electromagnetic interference shielding and thermal conductive performances, Chem. Eng. J. 380 (2020) 122475. https://doi.org/10.1016/j.cej.2019.122475

[69] M. Zhu, X. Yan, H. Xu, Y. Xu, L. Kong, Highly conductive and flexible bilayered MXene/cellulose paper sheet for efficient electromagnetic interference shielding applications, Ceram. Int. 47 (2021) 17234-17244. https://doi.org/10.1016/j.ceramint.2021.03.034

Chapter 3

MXenes for Nanophotonics

Sabahat Urossha[1] and S.S. Ali[1*]

[1]School of Physical Sciences, University of the Punjab, Lahore 54590, Pakistan

shahbaz.sps@pu.edu.pk

Abstract

Considerable attention has been paid to Mxenes which is a novel group of two-dimensional materials. Exceptional 2D layered microstructures make this family quite attractive for potential applications in a number of photoelectric applications along with their nonlinear optical and short electronic transport characteristics. This 2D group of materials comprises of transition metal carbides, carbonitrides, and nitrides having formula $M_{n+1}X_n$ (where M = Sc, Ti, Zr, Hf, V, Nb, Ta, Cr, Mo, etc., and X = C and/or N). Owing to some unmatched features of Mxenes like hydrophilic surfaces and high conductivities (~6000–8000 S/cm), these can be implemented in potential applications such as catalysis, energy storage, and field effect transistors etc.

Keywords

MXene, 2D Materials, Non-Linear Optical Behavior, Optoelectronic Properties, Photodetectors, Light Emitting Diodes

Contents

1.   MXenes -An introduction and as a 2D Material

MXene is the new branch of 2-dimensional nanomaterials, discovered by Gogotsi and Barsoum in 2011 who demonstrated that the HF can remove the Al atom from Ti_3AlC_2 which is one of the MAX type materials and Ti_3C_2 left behind, that provides the discovery of MXene later [1]. The ternary metal carbide/ nitride known as MXene, basically are members of transition metal carbides and exist in 2 dimensions, nitrides and carbon nitrides [2-3]. MXenes have gained a lot of interest owing to their exceptional performance with a variety of applications in different fields. The 2D MXenes are formed by omitting 'A' layer from the MAX phases, 2-dimensional MXene having 'ene' as a suffix to represent that it is almost identical to graphene [4].

MXenes usually have chemical formula $M_{n+1}AX_n$, where M denotes metals that change over time, A denotes the elements of A-group i.e. group 13 and 14 and the value of n is in the range 1 to 3, as shown in Fig. 1. In structure of MAX phase, A is comparatively having weakened bond than the MX bonding [5-6]. This makes them chemically more reactive such as at high temperature they may decompose partially as shown in the following equation [7].

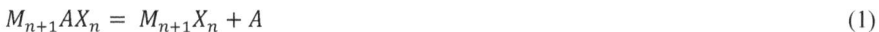

$$M_{n+1}AX_n \;=\; M_{n+1}X_n + A \tag{1}$$

The MXene have the general formula $M_{n+1}X_nT_x$, this chemical formula is for the functionalized MXene where n is from 1 to 3, here the early transition metals are represented by M, X represents the C or N as well as T_x represents surface termination like hydroxyl, oxygen and fluorine functional groups [8]. These termination groups can alter the properties of MXenes [9-12].

MXenes have broad application in various fields because of their fascinating electronic and optical properties, due to which they have wide applications such as it can be used in optoelectronic devices like photodetectors, photovoltaic, electronic, sensors, and in light emitting diodes (LEDs), etc. [13-15]. The MXene's properties may be adjusted by altering the surface functional group to meet desired requirements.

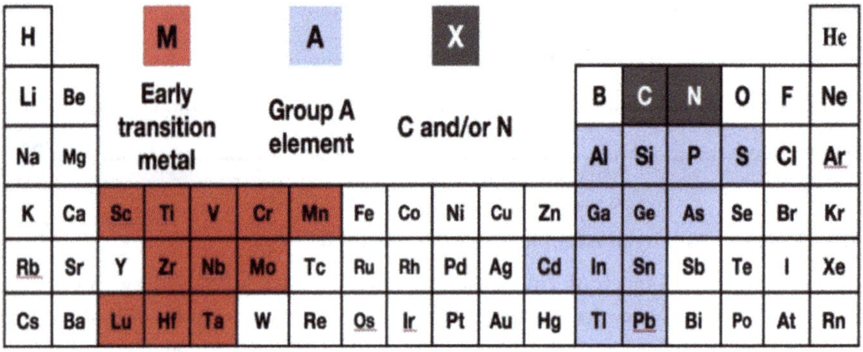

H		M		A		X											He
Li	Be	Early transition metal		Group A element		C and/or N						B	C	N	O	F	Ne
Na	Mg											Al	Si	P	S	Cl	Ar
K	Ca	Sc	Ti	V	Cr	Mn	Fe	Co	Ni	Cu	Zn	Ga	Ge	As	Se	Br	Kr
Rb	Sr	Y	Zr	Nb	Mo	Tc	Ru	Rh	Pd	Ag	Cd	In	Sn	Sb	Te	I	Xe
Cs	Ba	Lu	Hf	Ta	W	Re	Os	Ir	Pt	Au	Hg	Tl	Pb	Bi	Po	At	Rn

Figure 1.   MAX phase-forming elements. Reprinted from [M. Magnuson, M. Mattesini, Thin Solid Films 621 (2017) 108–130] with the permission of Elsevier publishing.

2.   Types of MXene

There are different types of MXene, but the semi-conductive type of MXenes are rare but essential. Semiconductive MXenes have more applications in optoelectronic devices [16-17]. These 2-dimensional materials are in high demands due to their different remarkable 'Optics' characteristics, such as significant optical nonlinearities, broad optical responsiveness and strong light matter interaction [18-19]. $Ti_3C_2T_x$ is one of the most important 2-dimensional MXenes. It has great potential in a variety of areas like in electronics, sensors that can detect EM interference and can be used in energy storage [20]. $Ti_3C_2T_x$ is by far the most investigated MXene, Researchers are making rapid progress on $Ti_3C_2T_x$ nanomaterial by taking use of its linear optical features [21]. $Ti_3C_2T_x$ exhibits a nonlinear absorption coefficient of $10^{-21} m^2/V^2$ that is equal to or higher than that of other 2D materials, synthesized by Zhang et al. [22]. At the moment, around 70 different kinds of MXene have been identified, V_2CT_x is another 2D MXene type, it has gained a great deal of attention due to its remarkable chemical and physical properties. In addition, V_2CT_x nano-sheets could be greatly beneficial for the opto-electronic devices in future [23]. V_2CT_x is the novel addition into the MXene family, to examine non-linear optical characteristics of this type of MXene, Z-scan technology was inspected [24-26], in nanoseconds, using 532 nm laser pulses. The V_2CT_x nano-sheets with high non-linear absorption co-efficient were examined experimentally, having approximate band-gap of 2.30-2.50 eV. V_2CT_x compounds can now be implemented in non-linear photonics because of their distinctive non-liner optical features, which have been discussed in this research.

Semi-conductive MXenes have the potential to increase the scope of MXene's use in disciplines like photonics and electronics. In addition to having an adequate band-gap, scandium-based MXenes also have a strong photo activated carrying capacity, which leading to enhance the performance of photo-catalysis [27]. Using etchants like HF to eliminate 'A' layer from MAX phase is most common approach for synthesizing MXene nowadays. Moreover, unstable MAX phase tends to make it an ideal choice for MXenes based on scandium. First time synthesis of Sc_2CO_x was done by using sputtering. Photoluminescence and light absorption spectra were used to study a sample's optical characteristics. MXenes have diverse and tunable electrical structures. The majority of MXenes are regarded metallic or semi-metallic, although a few are semiconducting, such as Ti_2CO_2, Sc_2CO_2 Zr_2CO_2 and Hf_2CO_2 [28]. These six semi-conducting types of MXenes have band gap values from 0.24 to 1.8 eV, covering visible to mid-infrared spectrum. In addition, termination groups and transition metal atoms can all influence how MXenes alter their band structures. The electrical structure of the MXene system can be affected by terminals such as -OH, -F, and -Cl, which are capable to receiving a single electron from metal atoms, whereas oxygen can get two [29].

3.   Non-linear optical behavior of MXene

There are different types of MXene which have been discussed above, now we will discuss their nonlinear optical behavior. In the field of modern technology, nonlinear optics is a widely controversial topic for researchers. It is concerned with light-matter interaction in which the reaction of objects to light is non-linear in relation to strength of EM field. Saturable absorption, four-wave mixing, optical rectification, and other fascinating phenomena can be induced in laser technologies using nonlinear optical material [30-32].

3.1 $Ti_3C_2T_x$ MXene

HF was used to selectively etch the MAX phase Ti_3AlC_2 and then extract the MXene $Ti_3C_2T_X$ from this. Initially, Ti_3AlC_2 was etched for six hours at room temperature with 50 (wt) % HF solution. Vacuum-assisted filtering was used to collect the $Ti_3C_2T_X$ residue upon a PVDF membrane. $Ti_3C_2T_X$ was baked in a vacuum oven at 80°C for 24 hours to form clay-like structure. $Ti_3C_2T_X$ powder and polyvinyl alcohol (PVA) were used to form the saturable absorber device's thin-film structure. Initially, it took 24 hours for combining the $Ti_3C_2T_X$ powder, 10 milligrams of PVA powder, and 40 milliliter deionized water at an ambient temperature. The resultant solution was then ultrasonically treated for two hours to separate the $Ti_3C_2T_X$ particle agglomerates. Lastly, the saturable absorber device for D-shaped fiber structure was easily made from the solution. It took 48 hours for the 5 milliliters of the standard solution to dry out on a clean petri dish. Then, the dried thin film was scraped away. D-shape fiber structure deposited MXene solution with a polished depth of 63.6 μm and polished length of 1.4 millimeters as depicted in Fig. 2(a).

The graph in Fig. 2 (b) and (c) depicts the non-linear transmittance, which had been derived through using the given equation.

$$T = 1 - \frac{\Delta T}{1+\frac{I}{I_s}} - \alpha_{ns} \tag{2}$$

where

ΔT = tramsmitance

I = intensity dependent input

I_s = saturation intensity

α_{ns} = non-saturable loss

A MXene-film SA was used for calculations, which revealed that the modulation depth was 1.22%, the maximum saturable power density was 0.4 MW/cm², and the loss was 82%, as shown in Fig. 2(b). Surface roughness, impurities, and defect absorptions in the SA device caused the MXene-non-saturable MXene film's loss to be quite high which gave rise the higher scattering loss to structure thin film. As a result of this, they demonstrated the non-linear transmittance of MXene saturable absorber and produced a graphical design as depicted in Fig. 2(c). In comparison to some MXene-based SA devices, MXene-DS' modulation depth is 3 percent, that is higher than MXene based device [1,33]. A low saturable intensity (0.2 MW/cm²), which suggests a SA device capable of starting pulsed lasers at lower threshold power. In the MXene-DS, the non-saturable loss was smaller than on the MXene-film which is **58.5** percent. Adding a tunable filter to the cavity would allow the mode-locked produced to operate at a wide range of wavelengths; but, the EDF emission cross-section would limit this. The nonlinear optical responses of $Ti_3C_2T_x$ MXene were also demonstrated by Jiang et al. at various wavelengths in the near-IR [34].

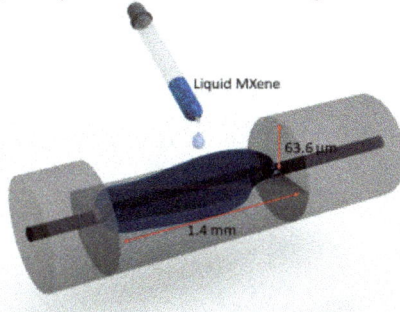

(a)

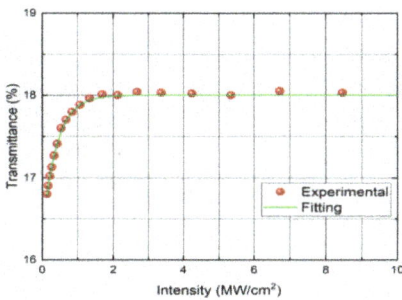

(b)

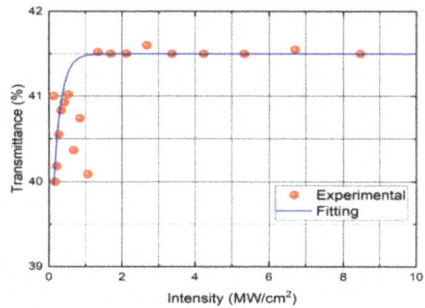

(c)

Figure 2.   *(a) Liquid MXene geometric structure placed on D-shaped fiber structure. (b) Non-linear transmittance of MXene-film and (c) non-linear transmittance of MXene-DS. Reprinted from [A. A. A. Jafry et al. Optics and Laser Technology 136 (2021) 106780] with the permission of Elsevier publishing.*

There are four ways in which nonlinear optical absorption can occur, as described in the following sections. The following equation represents the absorption coefficient (α) in terms of the effective non-linear absorption coefficient β:

$$\alpha = \alpha_0 + \beta_{eff} I \tag{3}$$

where,

α_0 = linear absorption-coefficient

I = laser intensity

The 3rd order non-linear optical susceptibility in terms of effective absorption coefficient is written as:

$$Im\chi^3 = \frac{2\varepsilon_0 c^2 n_0^2}{3\omega} \beta_{eff} \tag{4}$$

where

ε_0 = vacuum permeability

ω = angular frequency

c = speed of light and n_o represents the index of refraction ($n_o \approx 2$). In $Ti_3C_2T_x$, the β_{eff} is almost comparable to value of graphene-oxide, and in comparison, to the properties of black phosphorus by a factor of two-order as well as TMDC MoS_2 showing that substantial optical switching has been accomplished. We can measure its non-linear refractive index with a method known as closed aperture Z-scan.

Based on the following formula, a normalized transmittance was calculated:

$$T_{norm}(z) = \frac{1}{(1+x^2)^2 \Delta\varphi + \frac{4}{(1+x^2)^3 \Delta\varphi^2}} \cdot \frac{1-4x}{} \tag{5}$$

Where, x = z/z_o indicates the displacement of z-scan, $\Delta\varphi$ basically represents the phase change expressed as:

$$\Delta\varphi = \frac{2\pi n_2 I_o L_{eff}}{\lambda} \tag{6}$$

n = non-linear refractive index

L = physical thickness of sample and L_{eff} represents the effective thickness given by the following formula:

$$L_{eff} = \frac{(1 - e^{-\alpha_0 L})}{\alpha_0} \tag{7}$$

The real component of 3rd order nonlinear optical susceptibility correlated to non-linear refractive index expressed as:

$$Re\chi^3 = \frac{4\varepsilon_0 c n_0^2}{3} n_2 \tag{8}$$

It should be noted, even if the final results would be different depending on the experiment's design, however the calculated non-linear refractive-index (n) of $Ti_3C_2T_x$ was similar to those of graphene.

3.2 V_2CT_x MXene

V_2CT_x mono-sheets are a new addition to the MXene family, and their wideband and powerful non-linear optical responses have been described in different studies. Z-scan technology was used to examine their non-linear optical characteristics. According to experimental outcomes, the non-linear absorption coefficient of V_2CT_x was found to be large. To study non-linear optical properties of V_2CT_x, its mono-layer nano sheets were devised by selectively eliminating the Al element from the V_2AlC at room-temperature. Nanosecond Z-scan was performed to analyze non-linear optical characteristics of nano-sheets with a wavelength having values of 450–700 nm. The z-scan technique revealed that V_2CT_x nano-sheets possessed saturable absorption plenty of the periods because of bleaching the ground-state plasma. The findings of experiments exhibited V_2CT_x non-linear optical characteristics. The amount of V_2CT_x non-linear absorption varied with different values of wavelength.

3.2.1   Synthesis of V_2CT_x MXene

One gram powder (400 mesh) of V_2AlC was gently added to 50% of HF-solution and agitated for 90 hours at room temperature [35-36]. To prevent the V-Al bonds in V_2AlC from being damaged, the V-C bonds must be entirely broken in this reaction. The de-ionized H_2O was used to wash the precipitate until it reached a neutral pH ≥ 6. V_2CT_x was delaminated by shaking for 30 minutes in 25% TMAOH solution with 1 g of multilayer powder. Subsequent sedimentation at 3000 rpm removed the extra TMAOH from product. For 2 dimensional V_2CT_x MXene materials, freeze-drying has been used.

3.2.2 Characterization Results

Fig. 3(a) demonstrates the rough surface of V_2CT_x multilayers that is characterized by a shape which looks like an accordion. As depicted in Fig. 3(b,c), SEM and EDS analyses demonstrated that elements including V, C, O, and F were equally arranged in a one-layer V_2CT_x. The x-ray diffraction technique was utilized to analyze sample phase transitions after etching [(Fig. 3(d)]. Diffractions at $2\theta = 9.18°$ and $12.73°$ have been found as characteristic of V_2CT_x MXene just after etching and were related to the (002) crystal plane. V_2CT_x particles appear to have 2D layer-structure as seen in Fig. 3(e) showing TEM image of V_2CT_x. V_2CT_x-layers are visibly orientated along the (002) plane of crystal and calculated D-spacing is around 0.26 nanometer in Fig. 3(f). The FFT pattern in Fig. 3(g) exhibited a regular hexagonal lattice, showing that produced V_2CT_x has the exact hexagonal structure like that of the parental V_2Al. As shown in Fig. 4(a), the transmittance of V_2CT_x nano-sheets increased dramatically as the sample got closer to the laser beam focus (Z = 0), and a peak was seen, indicating that the material exhibited saturable absorber properties. Fig. 4(b) shows that the transmission enlarged initially when the sample gets closer to point of focus, showing SA property. The transmittance decreased when the sample left the focal point, showing that sample's saturable absorption characteristic had been transformed to reversed saturable absorption. Researchers observed reversed saturable absorption behavior when they used a significantly stronger laser pulse, non-linear optical characteristics were depicted in Fig. 4(a,c). The data in Fig. 4(d,f) is shown to manifest Z-scan analysis below 532 nm pulse operated at 3 various values of energy. One peak was found in the normalized transmission curve, showing that V_2CT_x nano-sheets possess SA characteristics. There was an increase in transmission when energy of incident-pulsation was 1380 uJ (Fig. 4(f)), which indicates a switch property. In addition, the data reveal a shift from saturable absorber to reversed saturable absorber. Fig. 4(g-o) shows that most of the experiments show the saturable absorber characteristic. Fig. 4(i) represents the highest saturable absorption at 650 nm and 750 uJ.

Following is an illustration of the non-linear optical characteristics' underlying physical method. MXene's natural property and the energy-dependent laser pulse were primary contributors to these findings. The most well-known hypothesis states that the material's non-linear susceptibility is caused by intra-band transitions, inter-band transitions, hot electron excitations, and thermal effects [4]. SA is a photon SA mechanism which prevails non-linear absorption at lowest pulse energies in V_2CT_x monolayer flake systems [37]. There was less V_2CT_x population at ground state due to a decrease in excited state V_2CT_x monolayer nano-sheet pumping under laser irradiation when sample brought closer to the laser point of focus. The ground-state bleach of plasmon band was the name given to this occurrence. SA was noticed because more light was allowed to flow through the sample, resulting in a greater transmittance [38].

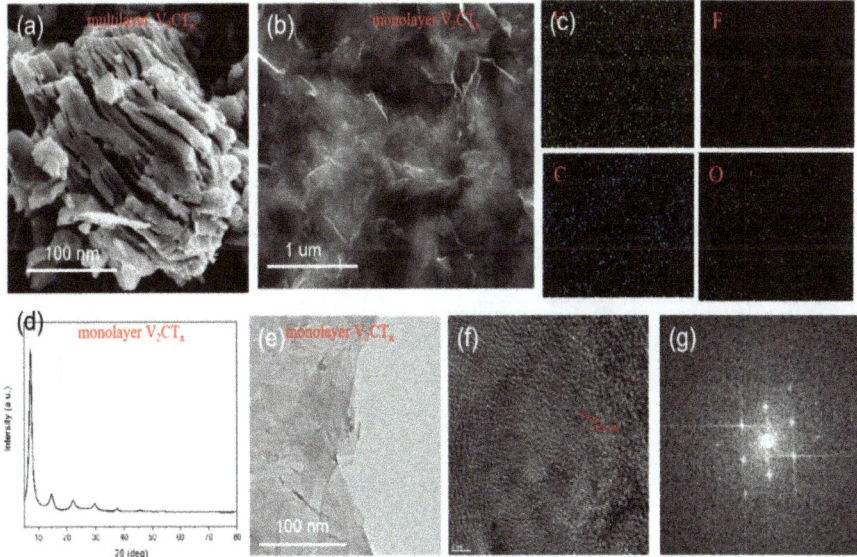

Figure 3.   (a,b) Scanning electron microscope images of multi and mono-layer V_2CT_x Nano-sheets. (c) Energy dispersive spectroscopy of mono-layer V_2CT_x. (d) X ray-diffraction structure of mono-layer V_2CT_x (e) transmission electron microscopy imaging of mono-layer V_2CT_x (f) V_2CT_x actual imaging with strong resolution TEM and FFT. Reprinted from [Q. He et al., Optik 247 (2021) 167629] with the permission of Elsevier publishing.

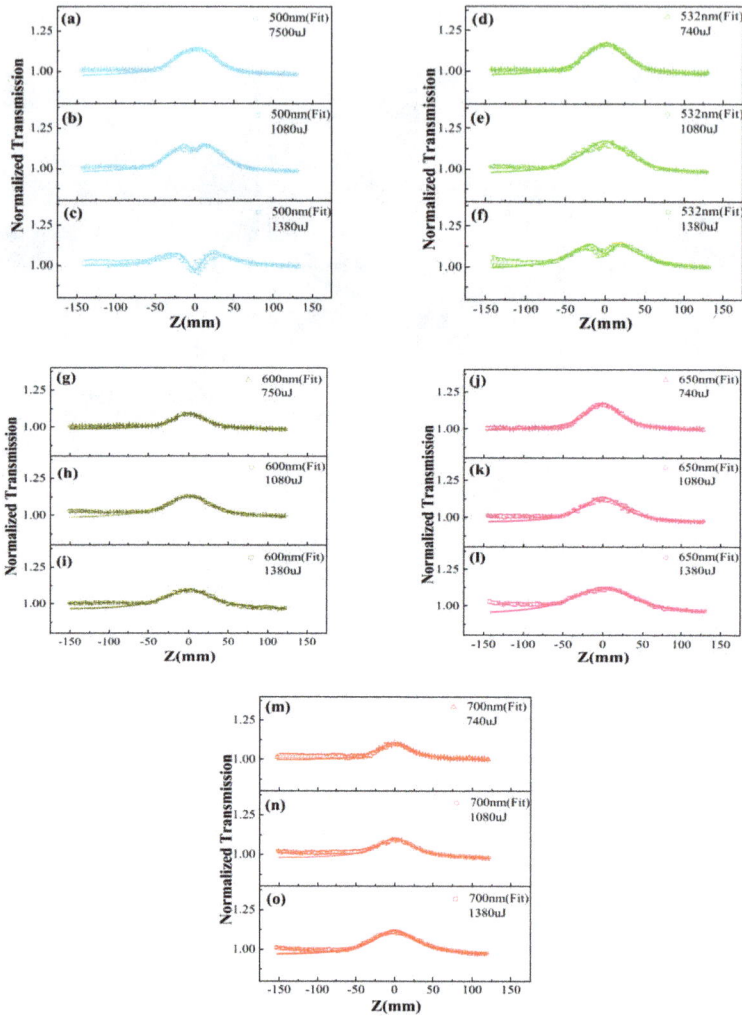

Figure 4. *Normalized transmission of V₂CTₓ Nano-sheets, z-scan dispersion position for wavelengths of 500 nm, 532 nm, 625 nm, 635 nm, and 700 nm in range of 750, 1080, and 1380 μJ laser energy respectively. Dots represent actual data, whereas lines represent theoretical model. Reprinted from [Q. He et al. Optik 247 (2021) 167629] with the permission of Elsevier publishing.*

Laser intensity and optical distance have a quantitative relationship as follows:

$$dI = -\alpha I dz \tag{9}$$

where

I = Intensity of laser

z = Optical distance and

α = Absorption co-efficient

In terms of total absorption-coefficient, it is represented as follows:

$$\alpha I = \frac{\alpha_o}{1+(\frac{I}{I_s})} + \beta I \tag{10}$$

Here, α_o is linear absorption-coefficient, I_s is saturable intensity and β represent positive non-linear absorption coefficient. In z-scan calculations, I is written as:

$$I = \frac{I_0}{1+\frac{z^2}{z_0^2}} \tag{11}$$

Equation 10 then becomes:

$$\alpha I = \frac{\alpha_o}{1+(\frac{\frac{I_0}{1+\frac{z^2}{z_0^2}}}{I_s})} + \beta \frac{I_0}{1+\frac{z^2}{z_0^2}} \tag{12}$$

From equations 9 and 12, normalized transmission is possible.

4.   Optical and Electronic Trends

4.1   Optical Properties

Many fascinating optical properties of MXenes have been discovered during last few years. Optical transparency, plasmonic activity, and effective photo-thermal transformation are only a few examples. The scientific world has benefited greatly from MXenes' capacity to interact with light in many ways. The most important factor in influencing a material's optical characteristics is its surface terminations. Berdiyorov, in 2016 studied the effect of material terminations upon optical characteristics of $Ti_3C_2T_2$ (T = F, O, OH) MXenes and did a comparison to MXene Ti_3C_2.

Because of unique optical properties and tune-ability, MXenes have a broad range of optical uses as a rising 2D material. Broadband optical transmittance of MXene films is estimated to be around 90% [39-40]. Additionally, the out-of-plane inter-band transitions and transmission valley about 750-780 nm, were thought to be caused by surface plasmon resonance (almost 800 nm) [41-43]. In particular, $Ti_3C_2T_x$'s optical transmittance in visible range was found to be 98 percent, which is greater than those of graphene and reduced graphene oxide, which were both experimentally tested [40].

MXenes' optical characteristics are influenced greatly by the surface functional groups that are incorporated during synthesis. For example, according to 1st principle, Ti_3C_2 with functional groups of fluorine and hydroxyl have lower absorption coefficients than pristine Ti₃C₂ and Ti₃C₂ with oxygen. But it turns out that surface functional groups have an influence on band-gaps of both metallic and semiconducting MXenes [41]. For example, the band-gap could be varied from 0.04 to 3.23 electron-volts or become a direct band-gap [42-45]. There are many other applications that could benefit from MXenes, including surface plasmon, terahertz (THz) technology and photoluminescence, but these are only a few examples. The photo response of semiconductors based on photoelectron technology was found to be superior to that of semiconductors based on photoelectron technology [46].

The optical characteristics of materials are essential for their use in electronic detectors and optical systems. The optical features can be derived from the complex dielectric function:

$$\varepsilon = \varepsilon_1(\omega) + i\varepsilon_2(\omega) \tag{13}$$

where

ω = given angular frequency, $\varepsilon_1(\omega)$ and $i\varepsilon_2(\omega)$ represent the real and imaginary parts. From the ε, various optical parameters including absorption coefficient α (ω), reflectance R (ω), and electron energy loss function L (ω) can be derived [47].

It is possible to calculate α (ω), by using the following expression.

$$\alpha\ (\omega) = \ \sqrt{2}(\omega)\left[\sqrt{\varepsilon_1(\omega)^2 + \ \varepsilon_2(\omega)^2} - \varepsilon_1(\omega)\right] \tag{14}$$

α (ω), reflectivity R and L (ω) are depicted in Fig. 5 [48].

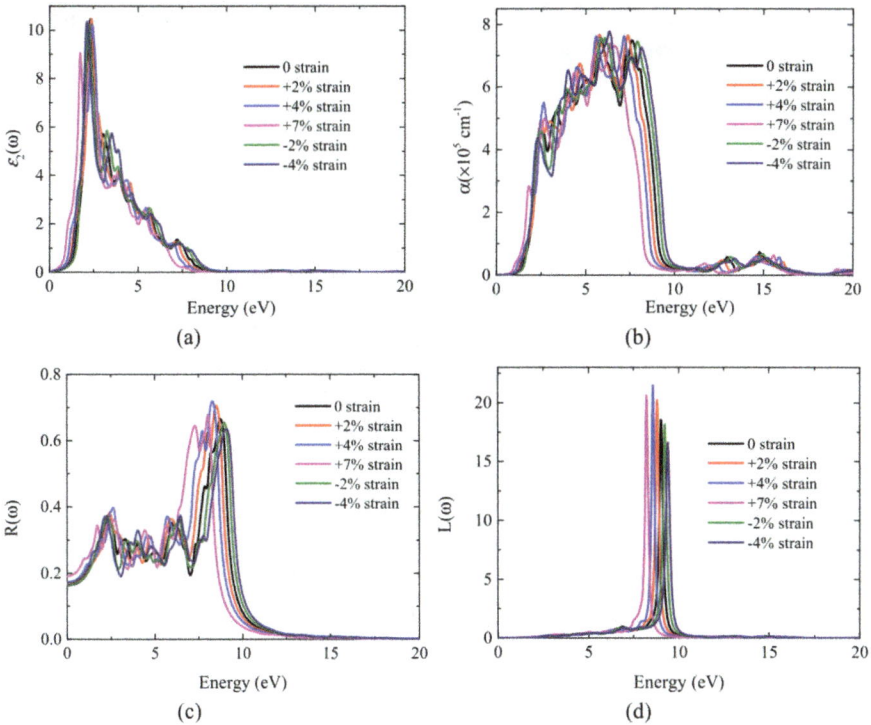

Figure 5.   *Optical properties of $Zr_2CO_2/MoSe_2$ (a) $\varepsilon_2(\omega)$ (b) $\alpha(\omega)$ (c) $R(\omega)$ (d) L (ω). Reprinted from [X. H. Li et al. Applied Surface Science 548 (2021) 149249] with the permission of Elsevier publishing.*

4.2   Electronic properties

MXene's conductivity is one of their important electronic properties. More pure MXenes and most of others with surface terminations display metallic behavior. The metallic conductivity of MXene has been increasing in recent years, even if the first discovered $Ti_3C_2T_x$ is by far the more conductive [49]. By regulating the surface terminations, novel $M_{n+1}X_n$ methods of synthesis are being developed that will result in the production of MXenes with greater conductivity. Furthermore, there is a lack of experimental verification in those investigations [50]. Cation or organic-molecule strong interaction is an alternative way to influence MXene conductivity, and it has the potential to raise the device's resistance by a factor of several orders of magnitude for stacked materials. Fig. 6 represents as the temperature rises above 500°C, the electrical conductivity of certain molybdenum based MXenes increases significantly [51-52].

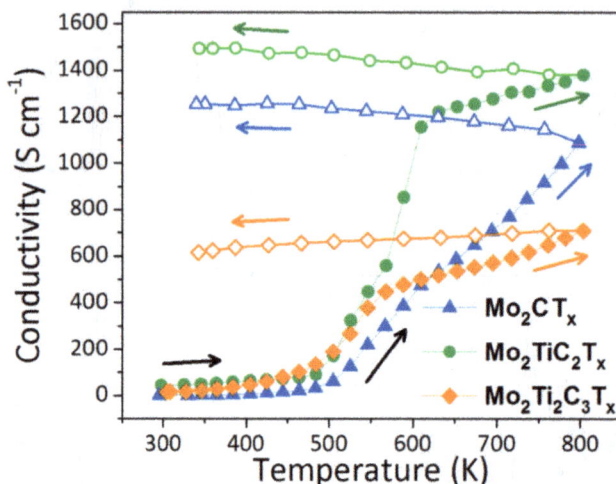

Figure 6.   Molybdenum based MXene with temperature sensitive electrical conductivity. The resistance decreases as the conductivity increases. Reprinted with permission from H. Kim et al., Chem. Mater. 29(15), 6472–6479 (2017). Copyright 2017 American Chemical Society.

Zhang et al. found that hydroxyl terminated MXenes have effectively free electron states adjacent to the surface [50]. These states occur into areas with largest positive charge and provide near-perfect electron transport routes, for example, $Ti_{n+1}C_nO_2$ had less electrical conductivity than MXenes ended with F or OH, $Ti_{n+1}C_nF_2$, and $Ti_{n+1}C_nOH_2$, in their tests. MXenes are all metallic, but a few of them became semi-conductors with band-gap relatively high when they are functionalized on the surface. The transition metal has a lower DOS at Fermi level because of electrons are transferred to electro-negative surface terminations [2,53].

5.   Theoretical outcomes

The calculations of density functional theory demonstrated that band-gaps of MXene could be controlled by changing surface functional group. The non-terminated MXene, for instance Ti_3C_2 behaves like a metallic conductor instead of this the surface functional group such as hydroxyl and fluorine are semiconductors having band-gap of 0.05 eV and 0.1 eV [1]. Thus, these materials could be used in a broad variety of scenarios, from FET to semiconductors, by adjusting their band-gaps [54-55].

By changing the elemental structure of M layer, the MXene's band gap can be varied. It is illustrated by Anasori et al. [46] that by replacing the top 2 layers of the Ti from $Ti_3C_2(OH)_2$ that is metal in nature with the molybdenum layer to generate the 2 dimensional double-layer transition metal carbides of $Mo_2TiC_2(OH)_2$ whereas the molybdenum atom exhibits conducting properties of MXene having 0.05 eV band gap.

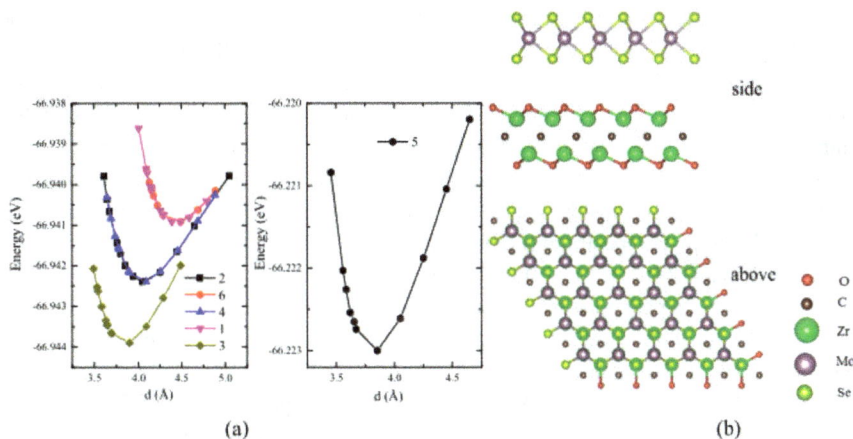

Figure 7.   (a) Ed curves for various stacking pattern (b) Structure of $Zr_2CO_2/MoSe_2$ hybrid bi-layer from sides and above. Reprinted from [X. H. Li et al. Applied Surface Science 548 (2021) 149249] with the permission of Elsevier publishing.

They upgraded the structures of Zr_2CO_2 and $MoSe_2$ monolayer. In the P63/mmc space group, Zr_2CO_2 and $MoSe_2$ are hexagonal. Zr_2CO_2 and $MoSe_2$ have lattice constants of 3.318 and 3.319 $A°$ respectively. Zr_2CO_2 MXene/$MoSe_2$ hybrid bilayer has a variety of stacking configurations. First, Mo is higher than Zr; second, Mo is higher than C; third, Mo is higher than O; fourth, Se is higher than O; fifth, Zr is higher than Se; sixth, Se is higher than C. In order to discover the most stable structure, the spacing between slabs must be varied in each arrangement. All of the stacking arrangements with varied distances between slabs were tuned for the best results. Except for stacking pattern 5, all binding energies are negative, indicating that stacking patterns other than stacking pattern 5 are energy viable. The energy-distance curves for each stacking arrangement are shown in Fig. 7(a). Overall, the least energy is accumulated in the stacking pattern 3 (Mo in the upper-most position, above the O atom). So, we adopted the $Zr_2CO_2/MoSe_2$ hybrid bilayer structure of stacking pattern 3 in all subsequent studies. Side and top views of $Zr_2CO_2/MoSe_2$ hybrid bilayer are shown in Fig. 7(b). The 2-dimensional heterostructure based optoelectronic devices require precise band alignment [48].

6.   Experimental outcomes

Fig. 8 compares the X-ray diffraction charts of Si substrate, ScC and ScO. Other than the peaks at (220), (311), (400), and (331) that correspond to the silicon substrate's crystal planes, the peaks at 67 and 61 can be attributed to scandium carbide's (220) and scandium oxide's (622) planes, correspondingly [56-57]. More information about the sample's composition can be found using the XPS measurement depicted in Fig. 9. Argon is used in XPS measurements to remove the surface layer. Scandium's sensitivity to oxygen makes it difficult to remove the element's contamination Fig. 10 shows the scandium, carbon, oxygen, and argon XPS spectra. Fig. 10(d) illustrates the use of argon in order to calibrate additional elements because of its stability and reactivity with other compounds. A typical value for Argon's 2p peak in silicon is at 242 eV [58]. The quantity and location of peaks can reveal details about sample's chemical components and bonds. It is clear from XPS spectrum of scandium as depicted in Fig. 10(a) that scandium possesses 2 types of bonds, since two independent sets of peaks are shown (a 2p3/2 peak and a 2p1/2 peak). The scandium-oxygen bond is represented by the 401 eV peak, while the scandium-carbon bond is represented by the 399 eV peak [58]. Scandium carbide and scandium oxide may exist as a result of this type of connection. Oxygen and carbon spectra add weight to the argument and round out the picture. The 281 eV peak of carbon in Fig. 10(b) implies the presence of a scandium-carbon bond, whereas the 284 eV peak demonstrates presence of a C-Sc-O link. MXene Sc$_2$C is functionalized by O-terminals, which means that a portion of it is pure [33]. Fig. 10(c) shows two oxygen bonds. The scandium oxide peak at 529 eV supports its presence, whereas the C-Sc-O bond peak at 531 eV confirms its existence [58]. It makes sense since Sc in a pristine MXene has a dangling link that attracts ions like F-, OH-, and O to form a functional group [59]. When exposed to air, the O- atoms in the sample will react with the scandium atoms to generate O-terminated MXene. Sc$_2$C, Sc$_2$CO, and scandium oxide makes up a bulk of the sample.

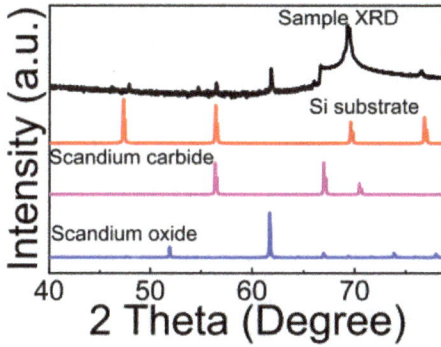

Figure 8.   X-ray diffraction trend of ScC without annealing. Reprinted from [Q. Chen, et al. Optik-International Journal for Light and Electron Optics 219 (2020) 165046] with the permission of Elsevier publishing.

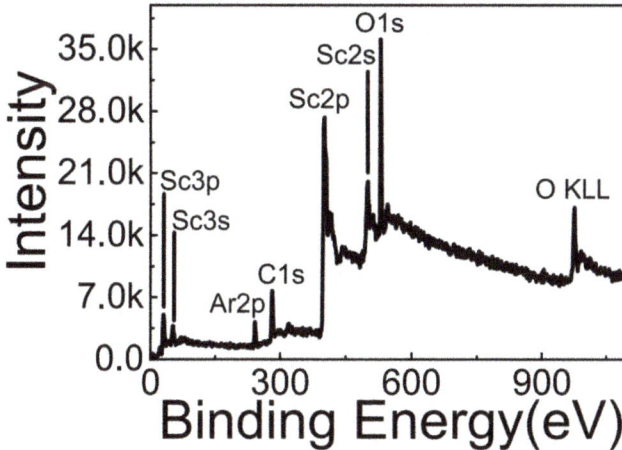

Figure 9.   X-ray photoelectron spectroscopy spectrum of un-annealed ScC. Reprinted from [Q. Chen, et al., Optik-International Journal for Light and Electron Optics 219 (2020) 165046] with the permission of Elsevier publishing.

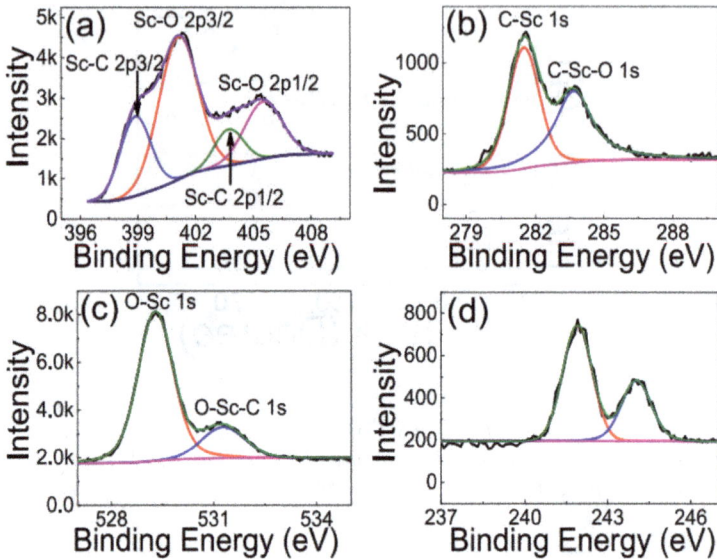

Figure 10.   X-ray photoelectron spectroscopy spectrum of (a) Sc (b) C (c) O and (d) Ar in an un-annealed scandium carbide sample. Reprinted from [Q. Chen, et al. Optik-International Journal for Light and Electron Optics 219 (2020) 165046] with the permission of Elsevier publishing.

7.   Device implementation

7.1   Saturable absorber

A saturable absorber is a material inserted within cavity of laser that enables initiation of pulsed lasers with pulse durations ranging from microseconds (10^{-6}) to femtoseconds (10^{-15}). This laser is a critical instrument for a wide range of applications, including material micro fabrication, skin therapy, and corrected Lasik in the scientific and industrial worlds [60-62]. In comparison to the high-intensity laser beam, the SA functions as a photon absorber. When optical output power of laser reaches predetermined value, group of electron moves from a ground state to the excited state. As a result of the Pauli exclusion principle, the saturable absorption mechanism is also unable to simultaneously hold two or more identical quantum states within the quantum system. An SA material must be able to endure significant optical damage, wide linear absorption, high saturation absorption capacity and also be easy to incorporate with laser arrangement [63]. There is no polarization dependence on the saturable absorber-based pulsed laser.

Recent Advances and Allied Applications of MXenes   Materials Research Forum LLC
Materials Research Foundations 155 (2024) 48-80   https://doi.org/10.21741/9781644902875-3

7.2   Photodetectors based on MXene

Optical devices that convert light signals into electrical signals, such as photodetectors, have been utilized widely in civilian and military applications, including satellite communication, missile guidance, remote sensing and night vision [64]. MXenes have been frequently employed in photodetectors because of their hydro-philicity, high conductivity, transparency, and ease of production. MXenes have a wide range of work functions, they can be utilized as an Ohmic or Schottky contact with different semiconductors, as a result of which they are utilized in wide range of photo detecting applications, including Schottky diodes, metal-semiconductor-metal photodetector, and photoconductor detector. Using transparent Ti_3C_2-based MXene electrode from water suspension on to gallium arsenide etched with photo-resist and pulled off with acetone, MXene-Semiconductor-MXene (M-S-MX) photodetectors that surpass ordinary Au electrodes are made [65].

Fig. 11 depicts the CdS core-Au/MXene photodetectors as were constructed. When a photo-response is negative, the potential in the electrode area is split into two separate components. The first one is the hot-electron injection from MXene to cadmium sulfide, while the other is the gold to CdS. The hot electrons inject from gold nanoparticles to CdS in the active area. Since MXene to cadmium sulfide transferred more electrons than from gold, a negative photo response was observed. The photoconductive effect was used for positive photo-response. The photo-thermal impact of MXene steadily reduces as the optical wavelength lowers, and the electron-hole pair rises. From gold to CdS, and subsequently from CdS to MXene, the electrons would be transferred. Positive photo-responses can be shown in this situation.

MXenes have 6000-10000 Scm^{-1} metallic conductivities. They could be used as conducting additives in photodetectors to increase photocurrent and photo response. F. Xia et al. built a UV photodetectors with zinc oxide nanoparticle-decorated BiOCl nano-sheet arrays on a PDMS substrate to improve the efficiency of charge carrier induced by light transport in composite photo detector [66]. To make the TiC/BiOCl/Cu sample, the BiOCl/Cu sample was spin-coated with the MXenes conduction layer. $Ti_3C_2T_x$ MXene as a conducting layer results in a 2 to 3 times magnitude increase in photocurrent and response time at 350 nm. $Ti_3C_2T_x$ MXene flakes may provide innovative and short paths for charge carriers to travel from BiOCl nano sheets to electrodes, resulting in a substantial increase in currents. Adding MXenes to a paired Zn_2GeO_4 nanowire network improved photo detection performance even further, as demonstrated by Guo et al. Further, the silica slide coated with Zn_2GeO_4 and MXene was easily dropped with MXene alcohol solution [67].

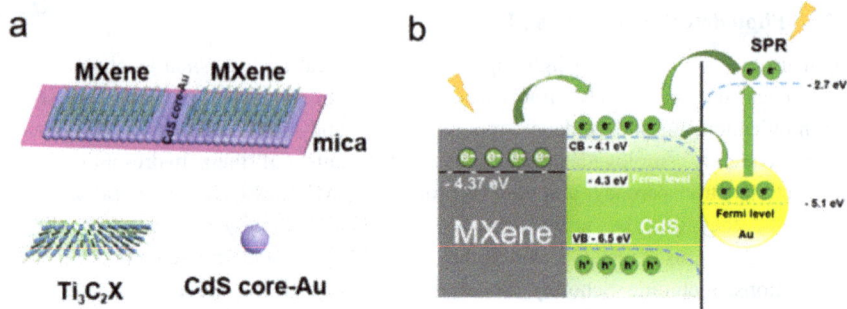

Figure 11. (a) Schematic of combined CdS core-Au/MXene photo-detector. (b) Diagram of the charge transfer mechanism as explained by energy-band engineering. Reprinted from [T. Jiang et al. Applied Surface Science 513 (2020) 145813] with the permission of Elsevier publishing.

Zn_2GeO_4/MXene had a photocurrent of 1.0 μA, which is substantially higher than Zn_2GeO_4. The Zn_2GeO_4/MXene rise/decay time is 1.16/0.39s, which is faster than the Zn_2GeO_4 rise/fall time. Under illumination at a wavelength of 254 nm, the Zn_2GeO_4/MXene hybrid nanostructures outperformed pure Zn_2GeO_4 in terms of optoelectronic performance, with a maximum response of 20.43 mA/W and an external quantum efficiency of 9.9 percent. MXene and Zn_2GeO_4 were said to work together in harmony to produce the better results. The sample's electron-hole generation is facilitated by the creation of unique semi conductive networks and massive Zn_2GeO_4 NWs contacts, interactions between the MXene layers. Titanium oxide is formed by the spontaneous oxidation of MXene flakes during delamination and storage in an open atmosphere. Thus, the MXene-titania composites may be employed for ultra-violet photo detection, with titanium oxide acting as an absorber of UV light and MXene as a conducting medium. Mochalin et al. reported that partially oxidized $Ti_3C_2T_x$ consisting of spontaneously generated titanium oxide nano particles reacted to ultra-violet exposure (254 nm) in a predictable manner [68].

7.3 Light emitting diodes

An electronic device which emits light when a current is passing through it is termed as LED. It is expected that in the future, flexible and wearable electronics will increasingly rely on LEDs that are highly energy-efficient and having sustainable solution manufacturing. The creation of a solution process-able and the ability to conduct electricity through transparent electrode, however, has hampered the advancement of these devices. MXenes are regarded among the most intriguing possibilities for innovative transparent electrode owing to excellent current flow, being optically transparent, and mechanical stability, see figure 12 [79].

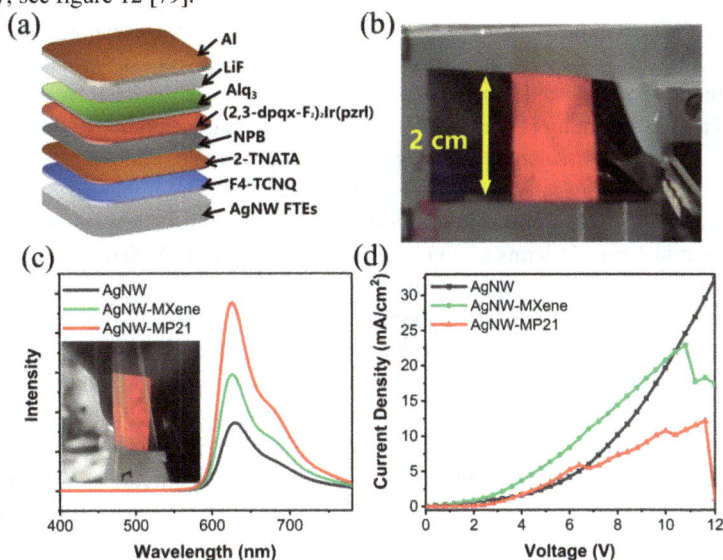

Figure 12.   (a) Structural design and (b) FOLEDs with an emission area of 2.8 cm2 based on AgNW-MP21 hybrid FTEs are shown in these images. Properties of OLED devices made with AgNW, MXene, or MP21 hybrid FTEs, like (c) Electroluminescence spectrum, (d) current density. Reprinted from [Shengchi Bai et al. Composites Part A. 139 (2020) 106088] with the permission of Elsevier publishing.

As illustrated in Fig. 12(a) and (b), 2.8 cm^2 red FOLEDs were constructed incorporating AgNW, AgNW-MXene and MXene-MP21 hybrid FTEs in order to evaluate influence of MXene and MP hybrids on electroluminescence. This red FOLED differs from prior studies [70, 71] by using commercially available materials for the rest of its variables. One of the many applications for flexible electronics is with the FOLEDs depicted in Fig. 12 (c), which may be bent in various ways. As shown in Fig. 12(c), the highest electro-luminescence peaks at 626 nanometers are essentially similar in shape, which indicate that these devices have weak or no micro holes [72]. FOLEDs by making use of AgNW, AgNW-MXene and AgNW-MP21 FTEs exhibit current density-voltage characteristics shown in Fig. 12(d), this is because of the higher leakage current and more surface roughness in the AgNW MXene hybrid FTE used in the FOLED'S [73].

7.4 Photovoltaic devices

In order to tackle the energy issue, photovoltaic systems could transfer sunlight into energy. Because of unique optical features, easiest device fabrication processes, solution process-able methodology, and band structure that can be tuned, two-dimensional PV devices have received a lot of attention [17,55,74]. New two-dimensional materials like MXenes offer enormous potential for solar applications because of their hydrophilic, highly conductive characteristics and transparency. MXenes, in particular, could be employed in organic and perovskite solar cells as transparent conducting electrodes or excellent electron and hole-transporting material. In terms of conductivity, the 2D $Ti_3C_2T_x$ MXene has a substantial advantage over solution-derived graphene [76]. A few layers of $Ti_3C_2T_x$ MXene were put onto the strongly doped n+ surface of the Si solar cell by He et al. [77]. For the np^+-Si side, phosphorus diffusion and micro-pyramidal surface texture are obtained by 70 nanometers, and boron alloying at 400 nanometers forms the nn^+-Si side, which is sandwiched between the two surfaces to increase inner absorption as well as reflecting of light. $Ti_3C_2T_x$ MXene deposition was made easier by forming an extremely thin n^+-Si before backscattering field to the deposition of $Ti_3C_2T_x$ MXene in order to reduce frictional resistance and recombination among $Ti_3C_2T_x$ MXene and Si. To further increase electrical contact among the $Ti_3C_2T_x$ MXene and Si device, the $Ti_3C_2T_x$ MXene linked with silicon device was thermally annealed under an inert environment, which reduced the contact resistance. Work functions were calculated through scanning Kelvin probe microscopy depending on the surface potential differences between 2 materials. Due to larger work function (0.18 eV) of $Ti_3C_2T_x$, at the interface of $Ti_3C_2T_x$ MXene/n^+-Si, an Ohmic contact is formed. The AM 1.5 G irradiation and subsequent quick thermal annealing under an Ar environment, demonstrates J-V properties of $Ti_3C_2T_x$ MXene/n^+ np^+ -Silicon based solar cell. With J_{sc} = 36.89 $mAcm^2$, V_{oc} = 0.55 V prior to quick thermal annealing procedure, the solar cell had an efficiency of 9.53 percent. Power conversion efficiency of 11.5% with J_{sc} = 36.70 $mAcm^{-2}$ and V_{oc} = 0.54 V was produced by annealing the device at 300°C. The annealing of $Ti_3C_2T_x$ MXene film at 300 °C resulted in a 20% increase in FF, which in turn resulted in a better power conversion efficiency, which can be attributed to the increase in conductivity of the film. This study showed that MXenes might be used as electrodes in the development of high-performance photovoltaic devices.

Materials Research Forum LLC
https://doi.org/10.21741/9781644902875-3

8.   Future perspectives and challenges

The optical and electrical features of MXenes are unique. Using MXenes as strong-performance opaque conductive electrodes in future transportable and worn electronics that are lightweight, transparent, adaptable, and mechanical stable is a great option because of its advanced optical and electronic features. Further transparent conducting materials like ITO, CVD-graphene, and RGO are depicted in Fig. 13.

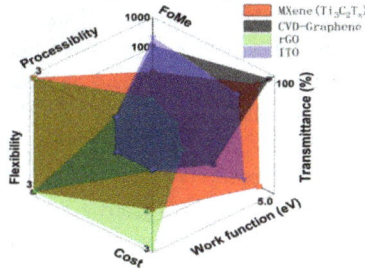

Figure 13.   MXene's comparison with transparent conducting materials like ITO, CVD-graphene and RGO. Reprinted from [X. Zhang, J. Shao, C. Yan et al. Materials and Design 200 (2021) 109452] with the permission of Elsevier publishing.

Different materials can be rated on their performance based on factors such as cost, adaptability, process ability and transmittance. Greater values of transmittance can be found in ITO. Flexibility is hindered by its brittle nature and high-temperature processing. Another drawback to ITO is its prohibitive cost. The promising new transparent conductive materials are CVD graphene and reduced graphene oxide. However, there are clear drawbacks to both of them. With their balanced performance across all six criteria, MXenes show significant promise as a material for the future generation of portable and wearable electronics. Apart from being a conducting additive, MXenes may also serve as an electron and hole-collection buffer material in optoelectronic applications because of their good qualities. MXenes are suitable for organic and perovskite optoelectronic devices manufacturing, as well as flexible electronics, due to their superior and modifying properties that are combining with a lowest temperature and solution process-able method. As a result, MXenes show great promise as a material for optoelectronic devices [78-84].

Despite MXenes' current success, there are still a number of unsolved issues. Only a few MXenes have been used in optoelectronic experiments. Metallic $Ti_3C_2T_x$ is the primary focus of most recent studies. New optoelectronic applications are planned for several semiconducting MXenes that will be produced. It is thought that direct band gap MXenes can make an important contribution to optoelectronics.

Conclusion

We discussed the distinct optoelectronic properties of MXene and recent advancements in their optoelectronic applications. Terminal groups, chemical intercalation, eliminating processes, transition metal, preparation techniques and size can all be used to alter MXene optical characteristics. MXenes' electrical properties are likewise exceptional, and their electronic structures can be adapted to fit specific applications. Metallic MXenes are thought to have a variety of options of work functions that are closely linked to the termination group. MXenes' tunable optoelectronic properties permit them to play a variety of functions in optical and electronics application, including transparent conductive electrodes, agents that conduct electricity and layers for charge transport. To summarize, we conclude that MXenes have an encouraging future and may emerge as an emergent alternative for optoelectronics just after overcoming obstacles in preparation and uses of these materials. Precursors, etching and delamination processes, among other things, are still restricting the manufacturing of large-sided, high-quality MXene grains. For the understanding of their characteristics and uses, a combination of theoretical and experimental work is required. For example, theoretical investigations of electrical and optical properties could successfully lead experimental work. Long-term air stability is still a problem for MXenes-based optoelectronic devices, preventing their widespread use0. Thus, stable MXenes and device packages will be needed in the future. If the aforementioned issues can be addressed, MXenes have the potential to be competitive materials for optoelectronic devices in general.

References

[1] M. Naguib, M. Kurtoglu, V. Presser, J. Lu, J. Niu, M. Heon, L. Hultman, Y. Gogotsi, M.W. Barsoum, Two-dimensional nanocrystals produced by exfoliation of Ti3AlC2, Adv. Mater. 23 (2011) 4248-4253. https://doi.org/10.1002/adma.201102306

[2] K. Hantanasirisakul, Y. Gogotsi, Electronic and optical properties of 2D transition metal carbides and nitrides (MXenes), Adv. Mater. 30 (2018) 1804779. https://doi.org/10.1002/adma.201804779

[3] M. Naguib, Y. Gogotsi, Synthesis of two-dimensional materials by selective extraction, Acc. Chem. Res. 48 (2015) 128-135. https://doi.org/10.1021/ar500346b

[4] M.W. Barsoum, The MN+1AXN phases: A new class of solids: Thermodynamically stable nano laminates, Prog. Solid State Chem. 28 (2000) 201-281. https://doi.org/10.1016/S0079-6786(00)00006-6

[5] M. Kurtoglu, M. Naguib, Y. Gogotsi, M. W. Barsoum, First principles study of two-dimensional early transition metal carbides, MRS Commun. 2 (2012) 133-137. https://doi.org/10.1557/mrc.2012.25

[6] M. W. Barsoum, Physical properties of the MAX phases, in: K.H.J. Buschow, R.W. Cahn, M.C. Flemings, B. Ilschner, E.J. Kramer, S. Mahajan, P. Veyssi'ere (Eds.), Encyclopedia of Materials: Science and Technology, Oxford, Elsevier, 2006.

[7] X. Jiang, A.V. Kuklin, A. Baev, Y. Ge, H. Ågren, H. Zhang, P.N. Prasad, Two dimensional MXenes: From morphological to optical, electric, and magnetic properties and applications, Phys. Rep. 848 (2020) 1-58. https://doi.org/10.1016/j.physrep.2019.12.006

[8] T. Hu, J. Wang, H. Zhang, Z. Li, M. Hu, X. Wang, Vibrational properties of Ti3C2 and Ti3C2T2 (T = O, F, OH) mono sheets by first-principles calculations: A comparative study, Phys. Chem. Chem. Phys. 17 (2015) 9997-10003. https://doi.org/10.1039/C4CP05666C

[9] A. N. Enyashin, A. L. Ivanovskii, Two-dimensional titanium carbonitrides and their hydroxylated derivatives: Structural, electronic properties and stability of MXenes Ti3C2−xNx (OH)2 from DFTB calculations, J. Solid State Chem. 207 (2013) 42-48. https://doi.org/10.1016/j.jssc.2013.09.010

[10] H. Lashgari, M. R. Abolhassani, A. Boochani, S. M. Elahi, J. Khodadadi, Electronic and optical properties of 2D graphene-like compounds titanium carbides and nitrides: DFT calculations, Solid State Commun. 195 (2014) 61-69. https://doi.org/10.1016/j.ssc.2014.06.008

[11] M. Khazaei, A. Ranjbar, M. G. Asl, M. Arai, T. Sasaki, Y. Liang, S. Yunoki, Nearly free electron states in MXenes, Phys. Rev. B. 93 (2016) 1-10. https://doi.org/10.1103/PhysRevB.93.205125

[12] N. K. Chaudhari, H. Jin, B. Kim, D. S. Baek, S.H. Joo, K. Lee, MXene: An emerging two-dimensional material for future energy conversion and storage applications, J. Mater. Chem. A 5 (2017) 24564-24579. https://doi.org/10.1039/C7TA09094C

[13] Z. Kang, Y. Ma, X. Tan, M. Zhu, Z. Zheng, N. Liu, L. Li, Z. Zou, X. Jiang, T. Zhai, Y. Gao, MXene-silicon Van Der Waals heterostructures for high-speed self-driven

photodetectors, Adv. Electron. Mater. 3 (2017) 1700165.
https://doi.org/10.1002/aelm.201700165

[14] S. Ahn, T.-H. Han, K. Maleski, J. Song, Y.-H. Kim, M.-H. Park, H. Zhou, S. Yoo, Y. Gogotsi, T.-W. Lee, A 2D titanium carbide MXene flexible electrode for high efficiency light-emitting diodes, Adv. Mater. 32 (2020) 2000919. https://doi.org/10.1002/adma.202000919

[15] Y. Uğur, Ö. Ayberk, K.P. Nihan, A. Feridun, S. Cem, Vibrational and mechanical properties of single layer MXene structures: A first principles investigation, Nanotechnol. 27 (2016) 335702. https://doi.org/10.1088/0957-4484/27/33/335702

[16] M. Khazaei, M. Arai, T. Sasaki, C.-Y. Chung, N.S. Venkataramanan, M. Estili, Y. Sakka, Y. Kawazoe, Novel electronic and magnetic properties of two-dimensional transition metal carbides and nitrides, Adv. Func. Mater. 23 (2013) 17. https://doi.org/10.1002/adfm.201202502

[17] J.W. You, S.R. Bongu, Q. Bao, N.C. Panoiu, Nonlinear optical properties and applications of 2D materials: Theoretical and experimental aspects, Nanophotonics. 8 (2019) 63-97. https://doi.org/10.1515/nanoph-2018-0106

[18] S. Sharifi, M.F. Nazar, F. Rakhshanizadeh, S.A. Sangsefedi, A. Azarpour, Impact of amino acids, organic solvents and surfactants on azo-hydrazone tautomerism in Methyl Red: pectral-luminescent and nonlinear optical properties, Opt. Quantum Electron. 52 (2020) 98. https://doi.org/10.1007/s11082-020-2211-3

[19] A. A. A. Jafry, G. Krishnan, N. Kasim, N. F. Zulkipli, F. S. M. Samsamnun, R. Apsari, S. W. Harun, MXene Ti3C2Tx as a passive Q-switcher for erbium-doped fiber laser, Optical Fiber Technol. 58 (2020) 102289. https://doi.org/10.1016/j.yofte.2020.102289

[20] H. Ghanadan, M. Hoseini, A. Sazgarnia, S. Sharifi, Effect of ion pairs on nonlinear optical properties of crystal violet: Surfactants, nano-droplets, and in vitro culture conditions, J. Electron. Mater. 48 (2019) 7417-7426. https://doi.org/10.1007/s11664-019-07516-9

[21] Y. I. Jhon, J. Koo, B. Anasori, M. Seo, J. H. Lee, Y. Gogotsi, Y. M. Jhon, Metallic MXene saturable absorber for femtosecond mode-locked lasers, Adv. Mater. 29 (2017) 1702496. https://doi.org/10.1002/adma.201702496

[22] X. Zhang, M. Xue, X. Yang, Z. Wang, G. Luo, Z. Huang, X. Sui, C. Li, Preparation and tribological properties of Ti3C2(OH)2 nanosheets as additives in base oil, RSC Adv. 5 (2015) 2762-2767. https://doi.org/10.1039/C4RA13800G

[23] S. Sharifi, S.G. Salavatovna, A. Azarpour, F. Rakhshanizadeh, G. Zohuri, M.R. Sharifmoghadam, Optical properties of methyl orange-doped droplet and photodynamic therapy of Staphylococcus aureus, J. Fluoresc. 29 (2019) 1331-1341. https://doi.org/10.1007/s10895-019-02459-0

[24] M. Hoseini, A. Sazgarnia, S. Sharifi, Cell culture medium and nano-confined water on nonlinear optical properties of Congo Red, Opt. Quantum Electron. 51 (2019) 144. https://doi.org/10.1007/s11082-019-1865-1

[25] K. Xiong, P. Wang, G. Yang, Z. Liu, H. Zhang, S. Jin, X. Xu, Functional group effects on the photoelectronic properties of MXene (Sc2CT2, T = O, F, OH) and their possible photocatalytic activities, Sci. Report. 7 (2017) 1-8. https://doi.org/10.1038/s41598-016-0028-x

[26] P.A. Franken, A.E. Hill, C.W. Peters, G. Weinreich, Generation of optical harmonics, Phys. Rev. Lett. 7 (1961) 118. https://doi.org/10.1103/PhysRevLett.7.118

[27] J. Kerr, Phil. Mag. 50 (1985) 337 https://doi.org/10.1080/14786447508641302

[28] M. Bass, P. A. Franken, J. F. Ward, G. Weinreich, Optical rectification, Phys. Rev. Lett. 9 (1962) 446 https://doi.org/10.1103/PhysRevLett.9.446

[29] M. Naguib, J. Halim, J. Lu, K.M. Cook, L. Hultman, Y. Gogotsi, M.W. Barsoum, New two-dimensional niobium and vanadium carbides as promising materials for Li-ion batteries, J. Am. Chem. Soc. 135 (2013) 15966-15969. https://doi.org/10.1021/ja405735d

[30] M. Hoseini, A. Sazgarnia, S. Sharifi, Cell culture medium and nano-confined water on nonlinear optical properties of Congo Red, Opt. Quantum Electron. 51 (2019) 144. https://doi.org/10.1007/s11082-019-1865-1

[31] M. Naguib, V.N. Mochalin, M.W. Barsoum, Y. Gogotsi, 25th anniversary article: MXenes: A new family of two-dimensional materials, Adv. Mater. 26 (2014) 992-1005. https://doi.org/10.1002/adma.201304138

[32] M. Magnuson, M. Mattesini, Chemical bonding and electronic-structure in MAX phases as viewed by X-ray spectroscopy and density functional theory, Thin Solid Films. 621 (2017) 108-130. https://doi.org/10.1016/j.tsf.2016.11.005

[33] C.J. Zhang, S. Pinilla, N. McEvoy, C.P. Cullen, B. Anasori, E. Long, S.-H. Park, A. Ascaso, A. Shmeliov, D. Krishnan, Oxidation stability of colloidal two-dimensional titanium carbides (MXenes), Chem. Mater. 29 (2017) 4848. https://doi.org/10.1021/acs.chemmater.7b00745

[34] X.T. Jiang, S.X. Liu, W.Y. Liang, S.J. Luo, Z.L. He, Y.S.Q. Ge, H.D. Wang, R. Cao, F. Zhang, Q. Wen, J.Q. Li, Q.L. Bao, D.Y. Fan, H. Zhang, Broadband nonlinear photonics in few-layer MXene Ti_3C_2Tx (T = F, O, or OH), Laser Photonics Rev. 12 (2018) 10. https://doi.org/10.1002/lpor.201700229

[35] J. L. Hart, K. Hantanasirisakul, A. C. Lang, B. Anasori, D. Pinto, Y. Pivak, J. T. van Omme, S. J. May, Y. Gogotsi, and M. L. Taheri, Control of MXenes' electronic properties through termination and intercalation, Nat. Commun. 10 (2019) 522. https://doi.org/10.1038/s41467-018-08169-8

[36] H. Kim, B. Anasori, Y. Gogotsi, H. N. Alshareef, Thermoelectric Properties of Two-Dimensional Molybdenum-Based MXenes, Chem. Mater. 29, 6472 (2017). https://doi.org/10.1021/acs.chemmater.7b02056

[37] A. L. Ivanovskii, A.N. Enyashin, Graphene-like transition-metal nanocarbides and nanonitrides, Russ. Chem. Rev. 82 (2013) 735. https://doi.org/10.1070/RC2013v082n08ABEH004398

[38] J. Xu, J. Shim, J.-H. Park, S. Lee, MXene electrode for the integration of WSe_2 and MoS_2 field effect transistors, Adv. Funct. Mater. 26 (2016) 5328-5334. https://doi.org/10.1002/adfm.201600771

[39] X.-H. Zha, Q. Huang, J. He, H. He, J. Zhai, J.S. Francisco, S. Du, The thermal and electrical properties of the promising semiconductor MXene Hf_2CO_2, Sci. Rep. 6 (2016) 27971. https://doi.org/10.1038/srep27971

[40] B. Anasori, C. Shi, E.J. Moon, Y. Xie, C.A. Voigt, P.R.C. Kent, S.J. May, S.J.L. Billinge, M.W. Barsoum, Y. Gogotsi, Control of electronic properties of 2D carbides (MXenes) by manipulating their transition metal layers, Nanoscale Horiz. 1 (2016) 227-234. https://doi.org/10.1039/C5NH00125K

[41] X. H. Li, X. H. Cui, C. H. Xing, H. L. Cui, R. Z. Zhang, Strain-tunable electronic and optical properties of Zr_2CO_2 MXene and $MoSe_2$ van der Waals heterojunction: A first principles calculation, Appl. Surf. Sci. 548 (2021) 149249 https://doi.org/10.1016/j.apsusc.2021.149249

[42] C. F. Zhang, V. Nicolosi, Graphene and MXene-based transparent conductive electrodes and supercapacitors, Energy Storage Mater. 16 (2019) 102-125. https://doi.org/10.1016/j.ensm.2018.05.003

[43] K. Hantanasirisakul, M.Q. Zhao, P. Urbankowski, J. Halim, B. Anasori, S. Kota, C.E. Ren, M.W. Barsoum, Y. Gogotsi, Fabrication of Ti_3C_2Tx MXene transparent

Materials Research Forum LLC
https://doi.org/10.21741/9781644902875-3

thin films with tunable optoelectronic properties, Adv. Electron. Mater. 2 (2016) 1600050. https://doi.org/10.1002/aelm.201600050

[44] Y. Dong, S. Chertopalov, K. Maleski, B. Anasori, L. Hu, S. Bhattacharya, A.M. Rao, Y. Gogotsi, V.N. Mochalin, R. Podila, Saturable absorption in 2D Ti3C2 MXene thin films for passive photonic diodes, Adv. Mater. 30 (2018) 1705714. https://doi.org/10.1002/adma.201705714

[45] J.K. El-Demellawi, S. Lopatin, J. Yin, O.F. Mohammed, H.N. Alshareef, Tunable multipolar surface plasmons in 2D Ti3C2TX MXene flakes, ACS Nano. 12 (2018) 8485-8493. https://doi.org/10.1021/acsnano.8b04029

[46] B. Anasori, M.R. Lukatskaya, Y. Gogotsi, 2D metal carbides and nitrides (MXenes) for energy storage, Nat. Rev. Mater. 2 (2017) 16098. https://doi.org/10.1038/natrevmats.2016.98

[47] G.R. Berdiyorov, Optical properties of functionalized Ti3C2Tx (T = F, O, OH) MXene: First-principles calculations, AIP Adv. 6 (2016) 055105. https://doi.org/10.1063/1.4948799

[48] J.H. Liu, X. Kan, B. Amin, L.Y. Gan, Y. Zhao, Theoretical exploration of the potential applications of Sc-based MXenes, Chem. Chem. Phys. 19 (2017) 32253-32261. https://doi.org/10.1039/C7CP06224A

[49] M. Khazaei, A. Ranjbar, A. Masao, S. Yunoki, Topological insulators in the ordered double transition metals M2′M″C2 MXenes (M′=Mo, W; M″=Ti, Zr, Hf), Phys. Rev. B. 94 (2016) 125152. https://doi.org/10.1103/PhysRevB.94.125152

[50] C. Zhang, V. Nicolosi, Graphene and MXene-based transparent conductive electrodes and supercapacitors, Energy Storage Mater. 16 (2019) 102-125. https://doi.org/10.1016/j.ensm.2018.05.003

[51] L. Hong, R.F. Klie, S. Öğüt, First-principles study of size- and edge-dependent properties of MXene nanoribbons, Phys. Rev. B. 93 (2016), 115412. https://doi.org/10.1103/PhysRevB.93.115412

[52] D.B. Velusamy, J.K. El-Demellawi, A.M. El-Zohry, A. Giugni, S. Lopatin, M.N. Hedhili, A.E. Mansour, E.D. Fabrizio, O.F. Mohammed, H.N. Alshareef, MXenes for plasmonic photodetection, Adv. Mater. 31 (2019), 1807658. https://doi.org/10.1002/adma.201807658

[53] S. Juodkazis, V. Mizeikis, H. Misawa, Three-dimensional microfabrication of materials by femtosecond lasers for photonics applications, J. Appl. Phys. 106 (2009) 14. https://doi.org/10.1063/1.3216462

[54] S. Graf, G. Staupendahl, A. Kramer, F.A. Muller, High precision materials processing using a novel Q-switched CO2 laser, Opt. Lasers Eng. 66 (2015) 152-157. https://doi.org/10.1016/j.optlaseng.2014.09.007

[55] S. Shah, T.S. Alster, Laser treatment of dark skin: An updated review, Am. J. Clin. Dermatol. 11 (2010) 389-397. https://doi.org/10.2165/11538940-000000000-00000

[56] S.A. Hussain, Discovery of several new families of saturable absorbers for ultrashort pulsed laser systems, Sci. Rep. 9 (2019) 19910. https://doi.org/10.1038/s41598-019-56460-5

[57] W. Jin, L. Hu, Review on quasi one-dimensional CdSe nanomaterials: Synthesis and application in photodetectors, Nanomaterials. 9 (2019) 1359. https://doi.org/10.3390/nano9101359

[58] K. Montazeri, M. Currie, L. Verger, P. Dianat, M.W. Barsoum, B. Nabet, Beyond gold: Spin-coated Ti3C2-based MXene photodetectors, Adv. Mater. 31 (2019) 1903271. https://doi.org/10.1002/adma.201903271

[59] Z. Kang, Y. Ma, X. Tan, M. Zhu, Z. Zheng, N. Liu, L. Li, Z. Zou, X. Jiang, T. Zhai, Y. Gao, MXene-silicon van der Waals heterostructures for high-speed self-driven photodetectors, Adv. Electron. Mater. 3 (2017) 1700165. https://doi.org/10.1002/aelm.201700165

[60] S. Bai, X. Guo, T. Chen, Y. Zhang, X. Zhang, H. Yang, X. Zhao, Solution processed fabrication of silver nanowire-MXene@PEDOT: PSS

flexible transparent electrodes for flexible organic light-emitting diodes, Composites Part A. 139 (2020) 106088. https://doi.org/10.1016/j.compositesa.2020.106088

[61] T. Kim, S. Kang, J. Heo, S. Cho, J. W. Kim, A. Choe. Nanoparticle-enhanced silver-nanowire plasmonic electrodes for high-performance organic optoelectronic devices, Adv. Mater. 30 (2018) 180065928. https://doi.org/10.1002/adma.201800659

[62] X. Zeng, Q. Zhang, R. Yu, C. Lu, A new transparent conductor: silver nanowire film

buried at the surface of a transparent polymer, Adv. Mater. 22 (2010) 4484-8. https://doi.org/10.1002/adma.201001811

[63] Y. Duan, X. Wang, Y. Yang, D. Yang, P. Chen, Highly flexible peeled-off silver

nanowire transparent anode using in organic light-emitting devices, Appl. Surf. Sci.

351 (2015) 445-50.

[64] J. Chang, K. Chiang, H. Kang, W. Chi, J. Chang, C. Wu, A solution-processed molybdenum oxide treated silver nanowire network: a highly conductive transparent

conducting electrode with superior mechanical and hole injection properties,

Nanoscale. 7 (2015) 4572-9. https://doi.org/10.1039/C4NR06805J

[65] J. Jean, P.R. Brown, R.L. Jaffe, T. Buonassisi, V. Bulovic, Pathways for solar photovoltaics, Energy Environ. Sci. 8 (2015) 1200-1219. https://doi.org/10.1039/C4EE04073B

[66] F. Xia, X. Wei Sun, S. Chen, Alternating-current MXene polymer light-emitting diodes, Nanoscale. 11 (2019) 5231-5239. https://doi.org/10.1039/C8NR10461A

[67] S. Guo, S. Kang, S. Feng, W. Lu, MXene-enhanced deep ultraviolet photovoltaic performances of crossed Zn2GeO4 nanowires, J. Phys. Chem. C. 124 (2020) 4764-4771. https://doi.org/10.1021/acs.jpcc.0c01032

[68] S. Chertopalov, V.N. Mochalin, Environment-sensitive photoresponse of spontaneously partially oxidized Ti3C2 MXene thin films, ACS Nano. 12 (2018) 6109-6116. https://doi.org/10.1021/acsnano.8b02379

[69] M. Bernardi, M. Palummo, J.C. Grossman, Extraordinary sunlight absorption and one nanometer thick photovoltaics using two-dimensional monolayer materials, Nano Lett. 13 (2013) 3664-3670. https://doi.org/10.1021/nl401544y

[70] M.F. Bhopal, D.W. Lee, A. Rehman, S.H. Lee, Past and future of graphene/silicon heterojunction solar cells: A review, J. Mater. Chem. C. 5 (2017) 10701-10714. https://doi.org/10.1039/C7TC03060F

[71] D.H. Lien, J.S. Kang, M. Amani, K. Chen, M. Tosun, H.P. Wang, T. Roy, M.S. Eggleston, M.C. Wu, M. Dubey, S.C. Lee, J.H. He, A. Javey, Engineering light out coupling in 2D materials, Nano Lett. 15 (2015) 1356-1361. https://doi.org/10.1021/nl504632u

[72] A. Lipatov, H. Lu, M. Alhabeb, B. Anasori, A. Gruverman, Y. Gogotsi, A. Sinitskii, Elastic properties of 2D Ti3C2Tx MXene monolayers and bilayers, Sci. Adv. 4 (2018) eaat0491. https://doi.org/10.1126/sciadv.aat0491

[73] F. Shahzad, A. Iqbal, H. Kim, C.M. Koo, 2D transition metal carbides (MXenes): Applications as an electrically conducting material, Adv. Mater. 32 (202159). https://doi.org/10.1002/adma.202002159

[74] Y. Chen, Y.Y. Yue, S.R. Wang, N. Zhang, H.B. Sun, Graphene as a transparent and conductive electrode for organic optoelectronic devices, Adv. Electron. Mater. 5 (2019), 1900247. https://doi.org/10.1002/aelm.201900247

[75] H.A. Becerril, J. Mao, Z. Liu, R.M. Stoltenberg, Z. Bao, Y. Chen, Evaluation of solution processed reduced graphene oxide films as transparent conductors, ACS Nano. 2 (2008) 463-470. https://doi.org/10.1021/nn700375n

[76] Y.S. Woo, Transparent conductive electrodes based on graphene-related materials, Micromachines. 10 (2019) 1-27. https://doi.org/10.3390/mi10010013

[77] H.C. Fu, V. Ramalingam, H. Kim, C.H. Lin, X. Fang, H.N. Alshareef, J. H. He, MXene contacted silicon solar cells with 11.5% efficiency, Adv. Energy Mater. 9 (2019) 1900180. https://doi.org/10.1002/aenm.201900180

[78] M.M. Masis, S.D. Wolf, R.W. Robinson, J.W. Ager, C. Ballif, Transparent electrodes for efficient optoelectronics, Adv. Electron. Mater. 3 (2017) 1600529. https://doi.org/10.1002/aelm.201600529

[79] C. Zhang, B. Anasori, A.S. Ascaso, S.H. Park, N. McEvoy, A. Shmeliov, G.S. Duesberg, J.N. Coleman, Y. Gogotsi, V. Nicolosi, Transparent, flexible, and conductive 2D titanium carbide (MXene) films with high volumetric capacitance, Adv. Mater. 29 (2017) 1702678. https://doi.org/10.1002/adma.201702678

[80] Mdi. jade 6 (computer software), materials data, Livermore, CA, USA (2004).

[81] D.A. Permin, E.M. Gavrishchuk, O.N. Klyusik, S.V. Egorov, A.A. Sorokin, Self-propagating high-temperature synthesis of Sc2O3 nanopowders using different precursors, Adv. Powder Technol. 27 (2016) 2457-2461. https://doi.org/10.1016/j.apt.2016.08.025

[82] D. Briggs, X-ray photoelectron spectroscopy (XPS), Handbook of Adhesion: Second Edition, 2005.

[83] J.J. Feng, X.H. Li, T.C. Feng, Y.M. Wang, J. Liu, H. Zhang, An harmonic mode locked Er-doped fiber laser by the evanescent field-based MXene Ti3C2Tx (T = F, O, or OH) saturable absorber, Annalen Der Physik. 532 (2020) 7. https://doi.org/10.1002/andp.201900437

[84] R.Z. Zhang, H.L. Cui, X.H. Li, First-principles study of structural, electronic and optical properties of doped Ti2CF2 MXenes, Physica B. 561 (2019) 90-96. https://doi.org/10.1016/j.physb.2019.02.056

Recent Advances and Allied Applications of MXenes   Materials Research Forum LLC
Materials Research Foundations 155 (2024) 81-102   https://doi.org/10.21741/9781644902875-4

Chapter 4

Application of MXenes in Photodetectors

Anupam Patel and Rajendra Kumar Singh*

Ionic Liquid and Solid-State Ionics Lab, Department of Physics, Institute of Science, Banaras Hindu University, Varanasi, India

anupam3184@gmail.com, rajendrasingh.bhu@gmail.com*

Abstract

Photodetectors are semiconductor devices with photoelectric conversion functions that play a significant part in many fields, including UV radiation skin sensors, biological sensing, optical communication, biomedical sensing, etc. Scientists are interested in 2D materials because of their great physical, chemical, thermal, and flexibility qualities, nano-sized thickness, and outstanding electronic and optical capabilities. A novel class of 2D materials called MXenes contains carbon nitrides, transition metals carbides, and nitrides. MXenes have numerous advantages including excellent structural thermal stability, higher conductivity, excellent optical properties, etc. Therefore, MXenes are extensively attractive to the application fields of the photodetector.

Keywords

2D Materials, MXenes, Electronic and Optical Properties, Photodetector

Contents

1.   Introduction

Photodetectors are called photosensors (optical receivers) that convert the optical signal (incident light) to electrical signals (current or voltage). In daily life, photodetectors are commonly used in various fields including optical telecommunications, military, imaging, aviation, oximeters, UV radiation skin sensors, and biomedical sensing [1–4]. The two-dimensional (2D) materials field is explored since 2004 when 2D graphene material was successfully prepared from bulk graphite through exfoliation using the scotch tape method [5]. Graphene, a 2D monolayer of graphite (an allotrope of carbon), was discovered by Andre Geim and Kostya Novoselov. For innovative experiments in graphene 2D material, Sir Geim and Sir Novoselov were awarded the Noble Prize in 2010 in Physics [6]. Different kinds of 2D materials like graphene oxide (GO), reduced graphene oxide (RGO), graphene, graphitic carbon nitride (g-C_3N_4), hexagonal boron nitride, metal-organic frameworks (MOFs), MXenes, transition metal dichalcogenides (TMDs, e.g., MoS_2), layered double hydroxides, black phosphorous and covalent organic frameworks (COFs) have been revealed and applied in several fields such as electronics and optoelectronics, etc. 2D materials have recently received a lot of attention because of their unique properties such as excellent physical, chemical, and thermal properties, good flexibility, large surface area, nano-sized thickness, and excellent electronic and optical properties [7–9]. Due to their unique properties and day-by-day demands with broad application prospects in energy storage (e.g., batteries, supercapacitors), photodetectors, photocatalysis, transistor, sensor, topological insulators, and so on, various types of 2D materials are continuously being discovered as well as developed. The creation of photodetectors based on various kinds of 2D materials has been aggressively studied in the optoelectronic field. A variety of photodetectors based on 2D materials, including graphene, and TMDs (WS_2, $MoSe_2$, MoS_2, etc.), have been reported to enhance the performance in optoelectronic devices, according to the literature survey study. But these are associated with some limitations like zero band-gap, low absorbance, and low carrier mobility which limit their use in optoelectronic devices at ambient temperature [10,11]. However, Xenes (e.g., stanene, phosphorene, borophene, silicene, etc.), black phosphorous, boron nitride, perovskite, etc. 2D materials have several opportunities for application in the realm of optoelectronic devices [12–18]. The main classes of 2D materials are shown in Figure 1.

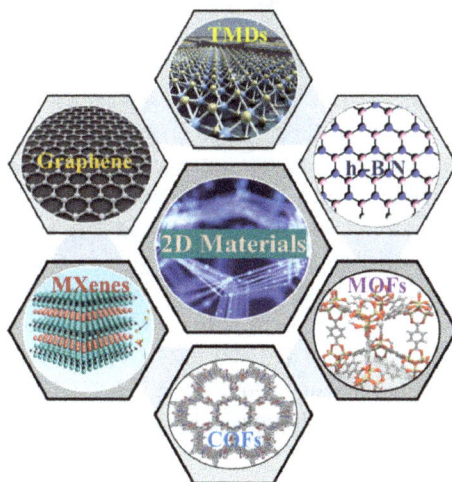

Figure 1. Schematic illustration of various categories of 2D materials.

Novel 2D MXene materials were revealed by Yury Gogotsi and Michel Barsoumin in 2011 [19]. The family of 2D transition metals known as MXene includes carbides and nitrides. Their general chemical formula is $M_{n+1}X_nT_x$, where M stands for a transition metal element such as Sc, Ti, Ta, Hf, Zr, V, Nb, or Cr, and n is one of one, two, or three. T_x indicates surface functional groups like O, OH, and F that are attached to the surface of MXene, and X denotes carbon (C) or nitrogen (N) atoms. Typically, MXenes are produced from layered precursors like MAX phases. The standard formulae for the MAX phase is $M_{n+1}AX_n$ (n = 1, 2, or 3), where M stands as an early transition metal, A represents an element from the IIIA or IVA group, such as Al, Ga, In, Si, Ge, Sn, Pb, P, As, etc., and X represents either C or N. Metal bands hold the atoms in the M and A layers of the MAX phase structure together, whereas the M-X band is made up of a combination of metallic, ionic, and covalent interactions [20–22]. M-A bands are easily broken due to weak force between M and A bands as compared to M-X bands. As a result, MAX phases like Ti_3SiC_2, Ti_2AlC, Ti_3AlC_2, Ti_2AlN, Ti_4AlN_3, Nb_4AlC_3, and so on are used to create MXenes by removing A

Materials Research Forum LLC
https://doi.org/10.21741/9781644902875-4

layer atoms. Based on n values (1,2 and 3), MAX phases can be formed by three different formulas like M_2AX or 211, M_3AX_2 or 312, and M_4AX_3 or 413. Also, MXenes can be assigned into three different forms such as M_2X, M_3X_2, and M_3X_4. There have been a number of MAX phases (155) recently documented in the literature. Therefore, a broader diversity of MXenes production may be possible given the increased number of MAX phases. According to literature surveys, out of 155 MAX phases, only more than twenty different types of MXenes have been reported [23]. As a result, numerous studies on MAX are being conducted to create new forms of MXenes. MXenes obtained via the MAX phase have numerous advantages such as excellent chemical and structural thermal stability, higher conductivity and hydrophilicity, large surface area, excellent optical properties, etc. Therefore, the use of MXenes is extensively attracted in many application fields like batteries, supercapacitors, thermoelectrics, sensors, field-effect transistors, and photodetectors [24–26]. In general, the $Ti_3C_2T_x$ and Ti_2CT_x types of MXenes are most widely used in various fields of applications [27,28]. However, there are just a few known uses of MXenes in photoreactors. Therefore, a lot of research on novel MXenes-based photodetectors can be expected in the near future. In this chapter, we have discussed various synthesis processes, characterization techniques, and properties of MXenes, as well as, focused on their application for photodetectors [29–31]. Figure 2 presents a synopsis of the synthesis, properties, and applications of MXenes.

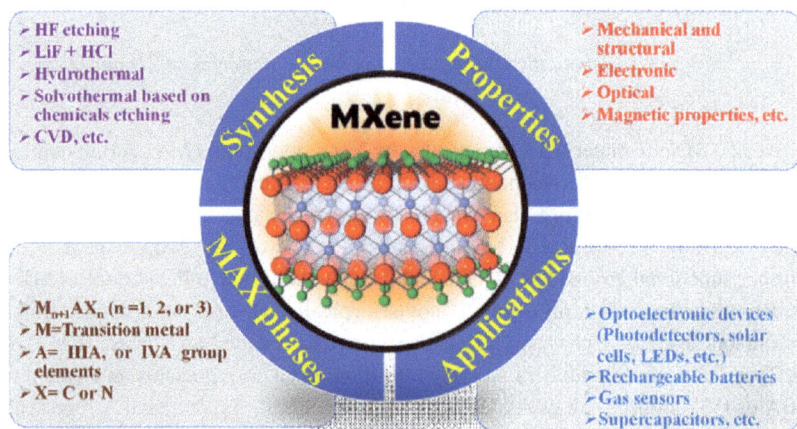

Figure 2. MAX phases and MXenes synthesis, properties, and their applications.

2. Preparation techniques of MXenes

2.1 Etching (HF etching) method

For the production of MXenes, numerous etching techniques have been devised and published. In 2011, the first MXene ($Ti_3C_2T_x$) was created by etching the Ti_3AlC_2 MAX phase using varied HF (hydrofluoric acid) concentrations. Naguib et al. described the synthesis of $Ti_3C_2T_x$ by etching Ti_3AlC_2 with HF solution, which is mentioned below. The chemical reactions of the MAX phase with HF during the synthesis of MXene have been obtained as follows: [19,32–34].

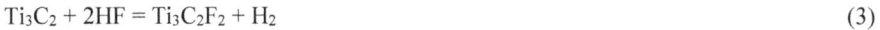

$$Ti_3AlC_2 + 3HF = AlF_3 + 3/2H_2 + Ti_3C \tag{1}$$

$$Ti_3C_2 + 2H_2O = Ti_3C_2(OH)_2 + H_2 \tag{2}$$

$$Ti_3C_2 + 2HF = Ti_3C_2F_2 + H_2 \tag{3}$$

From the chemical processes in equations 1, 2, and 3, it can be seen that the Ti_3AlC_2 reacts with HF to produce the Ti_3C by etching the Al layer atom from Ti_3AlC_2, and then Ti_3C interacts with H_2O and HF to form $Ti_3C_2(OH)_2$ and $Ti_3C_2F_2$, respectively. On the surface terminations of Ti3C, these reactions demonstrate the production of functional group like F^-, O^{2-}, or OH^- [35,36].

This process involved combining Ti_2AlC and TiC particles in a ball mill at a 1:1 molar ratio to create first the Ti_3AlC_2 MAX phase precursor. The material was heated at 1350°C for two hours in an argon (Ar) environment following ball milling. Finally, Ti_3AlC_2 was obtained at the end of the reaction. Now, 10 g of Ti_3AlC_2 was mixed in 100 mL HF (50% concentration) solution for 2 h at ambient temperature. The substance was filtered, then centrifuged to produce powders after being rinsed with deionized water. The structural and surface morphology analysis was investigated by XRD, Raman, XPS, and SEM to confirm MXenes preparation. Naguib et al. found that XRD patterns of both hydroxylated $Ti_3C_2(OH)_2$, and fluorinated $Ti_3C_2F_2$, structures were in good agreement with the simulated XRD patterns; The removal of Ti_3AlC_2's most intense diffraction peak at 39° and the good match between the simulated and actual XRD patterns for $Ti_3C_2(OH)_2$ give significant proof of its production. Prior to and during HF treatment, Ti_3AlC_2's Raman spectra display line broadening and spectrum shifts that are consistent with disintegration and match the broadened XRD patterns. XPS spectra taken before and after HF treatment revealed the presence of Ti-C and Ti-O bonds, according to Naguib et al. This indicates the creation of $Ti_3C_2(OH)_2$. Additionally, the surface termination groups may be seen in the XPS spectra.

According to the literature, various MAX phases have been used for the formation of MXenes using HF etching procedures. Various MAX phases like Ti_2AlC, Ti_3AlC_2, $TiNbAlC$, Ti_3AlCN, $(V_{0.5}Cr_{0.5})_3AlC_2$, $MoTiAlC_2$, etc. listed in Table-1, have been etched by HF to form the MXenes. When the MXene material is prepared by HF etching of the MAX phases materials, the quality of MXenes depends upon the following factors (a) the HF concentration, (b) the size of etched MAX phases material, (c) the etching time, and temperature of the etching during synthesis. If these factors are not properly established, the MXenes produced by the HF etching method will not be of good quality. The ideal HF concentration is critical for the etching process because when the concentration is low, the etching time and temperature are both too short. Therefore, the A atomic layer cannot be completely removed from MAX phases material to obtain the good quality of MXenes. However, when the HF concentration is high, the structure of the material that has been etched into the MAX phase is disrupted, and many flaws can be produced [22,23].

Table 1. M_2X, M_3X_2, and M_4X_3 types of MXenes synthesized by HF etchant.

Type	MXenes	Precursor	Etchant	Time [h]	Ref.
M_2X	Ti_2C	Ti_2AlC	10% HF	10	[24]
	Nb_2C Nb_2C	Nb_2AlC	50% HF 40% HF	90 48/72	[71,72]
	$TiNbC$	$TiNbAlC$	50% HF	28	[71]
	$(Ti_{0.5}V_{0.5})_2C$	$(Ti_{0.5}V_{0.5})_2AlC$	50% HF	19	[73]
M_3X_2	Ti_3C_2	Ti_3AlC_2	50% HF 10% HF 40% HF 49% HF	2 24 20 4-36	[19] [34,72,74]
	Ti_3CN	Ti_3AlCN	30% HF	18	[24]
	$(V_{0.5}Cr_{0.5})_3C_2$	$(V_{0.5}Cr_{0.5})_3AlC_2$	50%HF	69	[24]
	$(Ti_{0.5}V_{0.5})_3C_2$	$(Ti_{0.5}V_{0.5})_3AlC_2$	50%HF	18	[24]
	Mo_2TiC_2	Mo_2TiAlC_2	50% HF 48-51% HF	48 48	[75,76]
M_4X_3	V_4C_3	V_4AlC_3	40% HF	165	[73]
	Nb_4C_3	Nb_4AlC_3	48-51% HF 50% HF	96	[41]
	Ta_4C_3	Ta_4AlC_3	50% HF 50% HF	72	[24]
	$Mo_2Ti_2C_3$	$Mo_2Ti_2AlC_3$	50% HF	96	[76]
	$(Nb_{0.8}Ti_{0.2})_4C_3$	$(Nb_{0.8}Ti_{0.2})_4AlC_3$	50% HF	96	[77]

2.2   Non-HF etching methods

Additionally, the extensive use of HF, a toxic and strong corrosive acid, is hazardous to the human body (i.e., penetrates through the skin, bones, and muscle tissue), and the environment. Therefore, the researchers avoid the use of HF etchants to prepare the MXenes and other etchants have been established. Some other synthesis methods including plasma-enhanced pulse vapor deposition, non-HF etching (e.g., NH_4HF_2, and fluoride salts + HCl), hydrothermal, chemical vapor deposition (CVD), and so on that have been well known to minimize the use of HF.

The most popular method for creating MXenes materials among them is the non-HF etching method, which involves the use of a mixture of lithium fluoride and HCl. The chemical reaction of LiF and HCl is as follows:

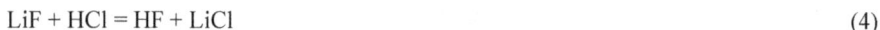

$$LiF + HCl = HF + LiCl \tag{4}$$

Form equation (4), it can be seen that LiF reacts with HCl to form the HF. This aqueous mixture (LiF + HCl) is less hazardous than that of direct use of HF etchant to synthesize the MXenes. Several MXenes synthesized by using LiF + HCl are listed in Table-2.

Table-2. Preparation of MXenes by using a mixture of the different weights of LiF and mol of HCl ratio solution.

Sample name	Synthesis approach (MAX)	Synthesis time & Temperature	Ref.
$Ti_3C_2T_x$	0.66g of LiF + 10 ml 6M HCl	45 h at 40 °C	[37]
$Ti_3C_2T_x$	1.0 g LiF + 20 ml 6.0M HCl	24 h at 35 °C	[75]
$Ti_3C_2T_x$	1.0 g of LiF + 20 ml 6M HCl	24 h at 35 °C	[78]
$Ti_3C_2T_x$	1.0 g LiF + 10 ml 9.0M HCl	24 h at 35 °C	[79]
$Ti_3C_2T_x$	1.0 g LiF + 25 ml 9.0M HCl	24 h at 45 °C	[80]
$Ti_3C_2T_x$	1.0 g LiF + 20 ml 9.0M HCl	24 h at 35 °C	[81]
$Ti_3C_2T_x$	1.6 g LiF + 20 ml 9.0M HCl	24 h at RT	[78]
$Ti_3C_2T_x$	1.6 g LiF + 20 ml 9.0M HCl	30 h at 50 °C	[82]
Ti_2C	4 g LiF + 40 ml 11.7M HCl	24 h at 35 °C	[83]
Ti_3CN	0.66g LiF + 10ml 6.0M HCl	12 h at 30 °C	[84]

Ghidiu et al. first introduced a lithium fluoride (LiF) and HCl solutions mixture in 2014, which was used to dissolve powders of the MAX phase (Ti_3AlC_2) in order to prepare the MXenes. Ti_3AlC_2 in LiF and HCl were mixed, and the mixture was then heated for 45 hours at 40°C. The clay-like substance was then produced using this technique after the residue had been washed [37]. In addition, different HCl concentrations and LiF/HCl molar ratios, along with other fluoride salts like NaF, KF, CsF, FeF$_3$, CaF$_2$, and NH$_4$F, have been utilized to create MXenes [38,39]. In the same year, Halim et al. also used ammonium hydrogen difluoride (NH$_4$HF$_2$) in place of HF etchant to eliminate the Al layer atoms from Ti$_3$AlC$_2$ to form the Ti$_3$C$_2$. During the etching method, the possible chemical reactions between the NH$_4$HF$_2$ and Ti$_3$AlC$_2$ are given below [40]:

$$Ti_3AlC_2 + 3NH_4HF_2 = (NH_4)_3AlF_6 + Ti_3C_2 + 3/2\ H_2 \qquad (5)$$

$$Ti_3C_2 + aNH_4HF_2 + bH_2O = (NH_3)c\ (NH_4)d\ Ti_3C_2(OH)_xF_y \qquad (6)$$

Both NH$_3$ and NH$_4^+$ molecules with Ti$_3$C$_2$ in the product, during the reaction, have the ability to intercalate in Ti$_3$C$_2$, raise interlayer spacing, and keep the distribution consistent. Whereas. the removal of some (NH$_4$)$_3$AlF$_6$ impurities is very difficult, which are produced at the same time during the reaction process. Therefore, the quality of Ti$_3$C$_2$ is not very good. In addition, the preparation of Ti$_3$C$_2$ was reported by Xie et al. in 2014, through using a mixture of NaOH and H$_2$SO$_4$ as etchants [41]. In this method, bulk Ti$_3$AlC$_2$ was immersed in the aqueous solution of NaOH and heated at 80 °C for 100 h. The Ti$_3$C$_2$ was then found after the resultant solution was treated in H$_2$SO$_4$ solution at 80 °C for two hours to remove additional Al atoms. This method is not effective due to taking too much time and the bad etching effect during synthesis. Additionally, different types of etchants like tetramethylammonium hydroxide (TMAOH), TMAOH + NH$_4$Cl, HCl, and ZnCl$_2$ have been reported to find the high quality of the MXenes [42–45].

2.3   Hydrothermal method

In the absence of HF, the hydrothermal approach successfully produces MXenes while avoiding the use of directly corrosive HF gas. Using ammonium fluoride (NH$_4$F) as an etching agent, Wang et al. disclosed a simple and environmentally friendly hydrothermal technique to manufacture MXene (Ti$_3$C$_2$T$_x$) in 2016 [46]. In this procedure, Wang et al. used magnetic stirring to dissolve 5 g of NH$_4$F in 60 mL of deionized water. After that, 0.5 g of Ti$_3$AlC$_2$ powder was well blended by stirring into the previously prepared solution. The aforementioned combination was additionally kept in an 80 mL stainless steel Teflon-lined autoclave for 24 hours of heating at 150°C. Following the conclusion of the natural cooling process, the precipitates were separated by centrifugation and repeatedly rinsed with deionized water and pure ethanol. Finally, samples were taken and dried in a vacuum oven for 12 hours at 60°C. Additionally, Li et al. (Li et al. 2018) prepared the fluorine-free high-quality of MXenes $Ti_3C_2T_x$ (T= –OH, –O) through a hydrothermal method [47]. MXenes quantum dots have been prepared by hydrothermal method. In 2017, Xue et al. synthesized the multilayer of $Ti_3C_2T_x$, firstly by using the HF to etch the Al layer atoms from the Ti_3AlC_2 and then Ti_3C_2QDs MXenes quantum dots were prepared from the multilayer $Ti_3C_2T_x$ [48].

3. Properties of MXenes

A brand-new class of 2D materials called MXenes has numerous excellent qualities, including mechanical, electrical, optical, and magnetic capabilities, etc.

3.1 Mechanical properties

The strong link between M-X (M-C or M-N) and the active functional groups O^{-2}, OH^-, F^- or Cl^- are coupled to the M (such as M-O, M-OH, M-F or M-Cl) on the surface of MXenes to improve considerably the desirable mechanical qualities. Generally, carbon (C) or nitrogen (N) atoms are connected to the transition metals representing the combination of covalent, ionic, and metallic bonding. The p orbitals of C or N are not full, and the d and s orbitals of transition metals are empty. Depending on the geometry of the transition metal carbide/nitride crystal structure, the covalent bonds in transition metal carbides and nitrides can vary. Through the d-orbitals, C or N do not create metallic connections between the transition metals. However, the contribution of the metallic bond between M-M is less compared to that of M–X bonds. Additionally, carbon or nitrogen deficiency regulates the impact of metallic and the impact of ionic/covalent M-X bonding. Strong mixed ionic/covalent M-O is added to MXenes structures ($M_{n+1}X_n$), which reduces the role of M-M interactions and increases the bond strength of the M-X bond [49,50]. According to a previous study on a simulation, the elastic parameters of MAX phases along with 2D components like CdS_2 have been proposed to be two times greater. While the bending rigidity is at its highest, the MAX phase's elastic characteristics are lower (2 to 3 times) than those of graphene. Additionally, compared to graphene, MXenes have excellent polymeric matrix contact abilities. The elastic constant of the $Ti_3C_2T_x$ monolayer has been determined experimentally via the nanoindentation procedure by using an AFM tip. In 2012, Murat Kurtoglu reported that MXenes' elastic constants are two times bigger than those of the comparable MAX phases [51]. Experimentally, Young's modulus (Y) of $Ti_3C_2T_x$ monolayer was determined to be 330± 30 GPa. The Y value of $Ti_3C_2T_x$ is calculated to be lower than the values of graphene (1000±100 GPa) and h-BN (870± 70 GPa) but larger than the graphene oxide (GO) = 210±20 GPa and MoS_2 = 270±100 GPa. MXenes are superior to other 2D materials and the most viable contender as support for composite materials due to their many good features, including increased bending stiffness, hydrophilicity, and high negative zeta potential (polymers, oxides, or carbon nanotubes). Numerous MXene-based composite polymers, such as $Ti_3C_2T_x$/PDDA and $Ti_3C_2T_x$/PVA, have been reported in experimental studies. It has been observed that the electrochemical capacitance, tensile strength, elastic moduli, thermal and electrical conductivities of the composite polymers enhanced in the presence of $Ti_3C_2T_x$ [52–55].

3.2   Structural properties

A brand-new family of 2D inorganic chemicals is called MXenes. Thick layers of transition metal carbides, nitrides, or carbonitrides makeup MXenes. Since the coordination number of transition metal ions is six. MXenes are believed to be a combination of hexagonal lattice structures with six-fold symmetry. Transition metals attached to surface functional groups create six chemical bonds to X atoms in MXenes as a result of the synthesis of M_2XO_2, M_2XF_2, and $M_2X(OH)_2$ [56].

3.3   Electronic properties

MXenes are highly electronic conducting materials, which is explained by the characteristics of M-layer, X-layer atoms, and terminal functional groups [57]. The electrical conductivity of MXenes has been found to be favorable to graphene and larger than GO and carbon nano [58]. According to DFT calculations, even without active functional groups, all Mxenes exhibit metal characteristics. However, some MXenes, such as $Ti_3C_2F_2$ and $Ti_3C_2(OH)_2$, exhibit semiconducting characteristics in the presence of terminal functional groups [19]. Additionally, MXenes show very high conductivity and carrier mobility which are highly influenced by the techniques used for material preparation and delamination. Also, MXenes with high anisotropy characteristics can be used in nanoelectronics devices [59,60]. In addition, some of MXenes (for example $M_2C_2O_2$, M= W, Mo, Cr) have been predicted as 2D topological insulators due to the effect of strong spin-orbit coupling. The $W_2C_2O_2$ type with strong spin-orbit and largest bandgap properties can be used in the photodetector field at ambient temperature. Furthermore, the different types of M'M''Xenes (M'M''C$_2$O$_2$, M'=W, Mo and M''= Ti, Zr, Hf) show 2D topological insulator properties. As a result, it is expected that MXenes will find widespread application in the field of photodetectors [61–63].

3.4   Optical properties

Regarding optics, MXenes with light emission, wide spectral response spectrum, photoluminescence, nonlinear light absorption, and light scattering properties perform optical activities. $Ti_3C_2T_x$ thin films and their composite materials are used to investigate the optical characteristics of MXenes. Thin films made of $Ti_3C_2T_x$ perform well both as an absorber and a transmitter. $Ti_3C_2T_x$ thin films have a high transmittance of 91% at 10 nm thickness and can absorb light in the UV region between 300 and 500 nm. The transmittance of the film can be changed with varying thicknesses and by using different intercalation precursors like urea, DMSO, hydrazine, etc. The transmittance percentage of $Ti_3C_2T_x$ film is decreased, when the urea, DMSO, and hydrazine are used. While the transmittance percentage increased from 75 to 92% when tetramethyl ammonium hydroxide was used [59,60,64,65]. MXenes are the most attractive in the field of several photonic and optoelectronic applications such as photothermal evaporation, conductive coating, optical switch, etc. due to their optical properties [48].

4. Application of MXenes in the field of photodetectors

Due to their unique qualities, such as wide interlayer spacing, hydrophilicity, strong surface reactivity, good biocompatibility, flexibility in the environment, safety, etc., MXenes have attained a great deal of popularity to date. The special characteristics of MXenes allow for applications in a variety of fields including optoelectronic devices, ferroelectricity, thermoelectricity, piezoelectricity, superconductivity, gas sensors, catalysis, EMI shielding (electromagnetic interference shielding), hydrogen storage, photocatalysis, triboelectric nanogenerators, lithium-ion, and lithium-sulfur batteries, supercapacitors, etc. MXenes with excellent electronic and optical properties are the most promising candidates for use in the field of photodetector applications. Recently, few photodetectors based on MXenes are reported. MXenes can therefore be used in a variety of photodetector applications. Some of the reported photodetectors based on MXenes are summarized in Table-3 with a preparation method and working mechanisms.

Zhou et al. (Zhou et al. 2019) fabricated a photodetector using MXenes (2D ScC_zOH) [66]. The 2D ScC_zOH was prepared by applying TMAOH etchant on the $ScAl_3C_3$ phase to remove the Al layer atoms. The ScC_zOH atomic structure, which contains OH- and carbon defects, was formed, as determined by the first principle DFT calculation. Also, ScC_zOH with 2.5 eV is shown as a direct bandgap semiconductor. The photodetector based on ScC_zOH has excellent photoresponse characteristics in the ultra-visible region. It can be expected that a photodetector based on ScC_zOH is extensively used in photocatalysis chemistry, visible light detector, and optoelectronic devices. In addition, photodetectors are prepared by using different Mxenes nanosheets like $Ti_3C_2T_x$, Ti_2CT_x, Mo_2CT_x, $NbCT_x$, and V_2CT_x. Among these, the photodetector based on Mo_2CT_x represented the best photoelectric performance [31].

Table-3. The photodetectors based on MXenes.

Materials	Preparation Working	Spectral Range [nm]	Responsivity [mA/W]	EQE [%]	Response times[ms]	Ref.
ScC_2OH	HF etching	280-600	125	43		[66]
Mo_2CT_x	HF etching	400-800	0.009	17		[31]
$Ti_3C_2T_z$/n-Si	LiF+9M HCl	405	26.95		0.84	[67]
Mo_2C/Graphene	CVD	400-1000	35.91		0.084	[68]
Mo_2C/MoS_2	CVD	405-1310	5×10^6			[85]
Ti_3C_2/Cu_2O	LiF+9M HCl	Visible			0.4	[69]
$CsPbBr_3/Ti_3C_2T_z$	LiF+6M HCl	Visible	44.9		18	[30]
InSe/ Ti_2CT_z	LiF+12M HCl	Visible-NIR	1.1×10^7		0.8	[70]
GaAs/ $Ti_3C_2T_z$	LiF+12M HCl	532	155	36		[86]

Kang et al. (Kang et al. 2017) also fabricated a van der Waals heterostructure photodetector by using Mxenes/semiconductor ($Ti_3C_2T_z$/n-Si). The design of vertical heterostructure can help to gain separation and transmission of photonic induced carriers, and $Ti_3C_2T_z$ via van der Waals form good Schottky contact with n-Si at room temperature. The photoresponsivity of the $Ti_3C_2T_z$/n-Si heterostructure photodetector is not high but its reaction and recovery times are fairly quick, and it continues to operate in self-driving mode [67]. As a result of their outstanding optoelectronic capabilities, MXenes/semiconductor vertical van der Waals heterostructures are being used increasingly in optoelectronics. They next created a new form of Mo_2C-graphene (Gr)/$Sb_2S_{0.42}Se_{2.58}$/TiO_2/fluorine-doped tin oxide vertical heterojunction based on the vertical heterostructure [68]. Additionally, Li et al. used a straightforward oil bath heating procedure to make a Ti_3C_2/Cu_2O heterojunction photoelectrochemical (PEC) photodetector to measure glucose concentration [69]. The device displays varying photoresponse properties depending on the glucose content. In a 1 μM concentration of glucose solution, it exhibits good stability. Their findings could lead to a new research direction for MXene applications in PEC photodetectors. In 2019, Deng et al. used $Ti_3C_2T_z$ as an electrode to build a fully sprayed, large-area flexible photodetector based on 2D $CsPbBr_3$/MXene [30]. Also, Yang et al. developed an InSe-based photodetector using Ti_2CT_z as an electrode based on a similar photodetector device designed using MXenes as the electrode [70].

Materials Research Forum LLC
https://doi.org/10.21741/9781644902875-4

According to research, 2D material photodetectors using MXene electrodes exhibit outstanding photoresponsivity and quick reaction times, which is significant for the development of new optoelectronic devices in the future.

Conclusion

MXenes will continue to encounter significant hurdles in photodetectors in the future. To begin with, MXene production methods are indeed limited, although chemical etching techniques can almost completely remove MAX phases with Al layers. Delaminated ultrasonic MXenes are still tiny layers rather than monolayer MXenes with functional groups. Furthermore, because existing methods for preparing MXenes are limited, it is necessary to create straightforward, affordable technology. Finally, because just a few MXenes are used in photodetectors, more developments into other MXenes in photodetectors are needed. MXenes, in general, offer a lot of appealing features and have been studied extensively. In several fields, these are commonly used. The researchers believe that MXenes have a lot of potential for enhancing the efficiency of current and emerging new high-performance photodetectors, as well as promoting the progress of 2D materials and photodetectors.

Acknowledgments

One of the authors (R.K. Singh) thankfully acknowledges the financial assistance by SERB, New Delhi and IOE, BHU to carry out the research work. A. Patel is thankful to UGC, New Delhi, for the award of SRF.

References

[1] C. Xie, F. Yan, Flexible photodetectors based on novel functional materials, Small. 13 (2017) 1-36. https://doi.org/10.1002/smll.201701822

[2] K. Zhang, L. Zhang, L. Han, L. Wang, Z. Chen, H. Xing, X. Chen, Recent progress and challenges based on two-dimensional material photodetectors, Nano Express. 2 (2021) 012001. https://doi.org/10.1088/2632-959X/abd45b

[3] T. Dong, J. Simões, Z. Yang, Flexible photodetector based on 2D Materials: Processing, architectures, and applications, Adv. Mater. Interfaces. 7 (2020) 1-18. https://doi.org/10.1002/admi.201901657

[4] L. Zheng, L. Zhongzhu, S. Guozhen, Photodetectors based on two dimensional materials, J. Semicond. 37 (2016). https://doi.org/10.1088/1674-4926/37/9/091001

[5] K.S. Novoselov, A.K. Geim, S.V. Morozov, D. Jiang, Y. Zhang, S.V. Dubonos, I.V. Grigorieva, and A.A. Firsov, Electric field effect in atomically thin carbon films, Science 306 (2016) 666-669. https://doi.org/10.1126/science.1102896

[6] A.K. Geim, K.S. Novoselov, The rise of graphene, Nat. Mater. 6 (2007) 183-191. https://doi.org/10.1038/nmat1849

[7] N. Ashraf, M. khan, A. Majid, M. Rafique, M.B. Tahir, A review of the interfacial properties of 2-D materials for energy storage and sensor applications, Chinese J. Phys. 66 (2020) 246-257. https://doi.org/10.1016/j.cjph.2020.03.035

[8] H. Zhu, J. Suhr, R. Ma, An overview of two-dimensional materials, J. Mater. 4 (2018) 81-82. https://doi.org/10.1016/j.jmat.2018.05.004

[9] H. Cui, Y. Guo, W. Ma, Z. Zhou, 2 D Materials for electrochemical energy storage: Design, preparation, and application, ChemSusChem. 13 (2020) 1155-1171. https://doi.org/10.1002/cssc.201903095

[10] W. Zheng, Y. Jiang, X. Hu, H. Li, Z. Zeng, X. Wang, A. Pan, Light emission properties of 2D transition metal dichalcogenides: Fundamentals and applications, Adv. Opt. Mater. 6 (2018) 1-29. https://doi.org/10.1002/adom.201800420

[11] H. Tian, M.L. Chin, S. Najmaei, Q. Guo, F. Xia, H. Wang, M. Dubey, Optoelectronic devices based on two-dimensional transition metal dichalcogenides, Nano Res. 9 (2016) 1543-1560. https://doi.org/10.1007/s12274-016-1034-9

[12] Y. Jiang, L. Miao, G. Jiang, Y. Chen, X. Qi, X.F. Jiang, H. Zhang, S. Wen, Broadband and enhanced nonlinear optical response of MoS2/graphene nanocomposites for ultrafast photonics applications, Sci. Rep. 5 (2015) 1-12. https://doi.org/10.1038/srep16372

[13] Y. Tan, X. Liu, Z. He, Y. Liu, M. Zhao, H. Zhang, F. Chen, Tuning of interlayer coupling in large-area graphene/WSe2 van der Waals heterostructure via ion irradiation: Optical evidence and photonic applications, ACS Photonics. 4 (2017) 1531-1538. https://doi.org/10.1021/acsphotonics.7b00296

[14] K.F. Mak, J. Shan, Photonics and optoelectronics of 2D semiconductor transition metal dichalcogenides, Nat. Photonics. 10 (2016) 216-226. https://doi.org/10.1038/nphoton.2015.282

[15] Y. Wang, F. Zhang, X. Tang, X. Chen, Y. Chen, W. Huang, Z. Liang, L. Wu, Y. Ge, Y. Song, J. Liu, D. Zhang, J. Li, H. Zhang, All-optical phosphorene phase modulator with enhanced stability under ambient conditions, Laser Photonics Rev. 12 (2018) 1-9. https://doi.org/10.1002/lpor.201800016

[16] S. Liu, Z. Li, Y. Ge, H. Wang, R. Yue, X. Jiang, J. Li, Q. Wen, H. Zhang, Graphene/phosphorene nano-heterojunction: Facile synthesis, nonlinear optics, and ultrafast photonics applications with enhanced performance, Photonics Res. 5 (2017) 662. https://doi.org/10.1364/PRJ.5.000662

[17] J. Zhao, D. Ma, C. Wang, Z. Guo, B. Zhang, J. Li, G. Nie, N. Xie, H. Zhang, Recent advances in anisotropic two-dimensional materials and device applications, Nano Res. 14 (2021) 897-919. https://doi.org/10.1007/s12274-020-3018-z

[18] J. Song, L. Xu, J. Li, J. Xue, Y. Dong, X. Li, H. Zeng, Monolayer and few-layer all-inorganic perovskites as a new family of two-dimensional semiconductors for printable optoelectronic devices, Adv. Mater. 28 (2016) 4861-4869. https://doi.org/10.1002/adma.201600225

[19] M. Naguib, M. Kurtoglu, V. Presser, J. Lu, J. Niu, M. Heon, L. Hultman, Y. Gogotsi, M.W. Barsoum, Two-dimensional nanocrystals produced by exfoliation of Ti3AlC2, Adv. Mater. 23 (2011) 4248-4253. https://doi.org/10.1002/adma.201102306

[20] Z. Zhang, Z. Cai, Y. Zhang, Y. Peng, Z. Wang, L. Xia, S. Ma, Z. Yin, R. Wang, Y. Cao, Z. Li, Y. Huang, The recent progress of MXene-Based microwave absorption materials, Carbon N. Y. 174 (2021) 484-499. https://doi.org/10.1016/j.carbon.2020.12.060

[21] B.M. Jun, S. Kim, J. Heo, C.M. Park, N. Her, M. Jang, Y. Huang, J. Han, Y. Yoon, Review of MXenes as new nanomaterials for energy storage/delivery and selected environmental applications, Nano Res. 12 (2019) 471-487. https://doi.org/10.1007/s12274-018-2225-3

[22] L. Verger, C. Xu, V. Natu, H.M. Cheng, W. Ren, M.W. Barsoum, Overview of the synthesis of MXenes and other ultrathin 2D transition metal carbides and nitrides, Curr. Opin. Solid State Mater. Sci. 23 (2019) 149-163. https://doi.org/10.1016/j.cossms.2019.02.001

[23] B. Wang, S. Zhong, P. Xu, H. Zhang, Booming development and present advances of two-dimensional MXenes for photodetectors, Chem. Eng. J. 403 (2021) 126336. https://doi.org/10.1016/j.cej.2020.126336

[24] M. Naguib, O. Mashtalir, J. Carle, V. Presser, J. Lu, L. Hultman, Y. Gogotsi, M.W. Barsoum, Two-dimensional transition metal carbides, ACS Nano. 6 (2012) 1322-1331. https://doi.org/10.1021/nn204153h

[25] C.E. Ren, M.Q. Zhao, T. Makaryan, J. Halim, M. Boota, S. Kota, B. Anasori, M.W. Barsoum, Y. Gogotsi, Porous two-dimensional transition metal carbide (MXene)

flakes for high-performance li-ion storage, ChemElectroChem. 3 (2016) 689-693. https://doi.org/10.1002/celc.201600059

[26] Y. Xie, M. Naguib, V.N. Mochalin, M.W. Barsoum, Y. Gogotsi, X. Yu, K.W. Nam, X.Q. Yang, A.I. Kolesnikov, P.R.C. Kent, Role of surface structure on li-ion energy storage capacity of two-dimensional transition-metal carbides, J. Am. Chem. Soc. 136 (2014) 6385-6394. https://doi.org/10.1021/ja501520b

[27] Z. Liu, H.N. Alshareef, MXenes for optoelectronic devices, Adv. Electron. Mater. 7 (2021) 1-28. https://doi.org/10.1002/aelm.202100295

[28] X. Jiang, S. Liu, W. Liang, S. Luo, Z. He, Y. Ge, H. Wang, R. Cao, F. Zhang, Q. Wen, J. Li, Q. Bao, D. Fan, H. Zhang, Broadband nonlinear photonics in few-layer MXene Ti3C2Tx (T = F, O, or OH), Laser Photonics Rev. 12 (2018) 1-10. https://doi.org/10.1002/lpor.201700229

[29] Y.I. Jhon, M. Seo, Y.M. Jhon, First-principles study of a MXene terahertz detector, Nanoscale. 10 (2018) 69-75. https://doi.org/10.1039/C7NR05351G

[30] W. Deng, H. Huang, H. Jin, W. Li, X. Chu, D. Xiong, W. Yan, F. Chun, M. Xie, C. Luo, L. Jin, C. Liu, H. Zhang, W. Deng, W. Yang, All-sprayed-processable, large-area, and flexible perovskite/MXene-based photodetector arrays for photocommunication, Adv. Opt. Mater. 7 (2019) 1-9. https://doi.org/10.1002/adom.201801521

[31] D.B. Velusamy, J.K. El-Demellawi, A.M. El-Zohry, A. Giugni, S. Lopatin, M.N. Hedhili, A.E. Mansour, E. Di Fabrizio, O.F. Mohammed, H.N. Alshareef, MXenes for plasmonic photodetection, Adv. Mater. 31 (2019) 1-10. https://doi.org/10.1002/adma.201807658

[32] T. Hu, Z. Li, M. Hu, J. Wang, Q. Hu, Q. Li, X. Wang, Chemical origin of termination-functionalized MXenes: Ti3C2T2 as a case study, J. Phys. Chem. C. 121 (2017) 19254-19261. https://doi.org/10.1021/acs.jpcc.7b05675

[33] M. Pogorielov, K. Smyrnova, S. Kyrylenko, O. Gogotsi, V. Zahorodna, A. Pogrebnjak, Mxenes-A new class of two-dimensional materials: Structure, properties and potential applications, Nanomaterials. 11 (2021) 1123412. https://doi.org/10.3390/nano11123412

[34] P. Srivastava, A. Mishra, H. Mizuseki, K.R. Lee, A.K. Singh, Mechanistic insight into the chemical exfoliation and functionalization of Ti3C2 MXene, ACS Appl. Mater. Interfaces. 8 (2016) 24256-24264. https://doi.org/10.1021/acsami.6b08413

[35] M. Khazaei, A. Ranjbar, M. Arai, T. Sasaki, S. Yunoki, Electronic properties and applications of MXenes: A theoretical review, J. Mater. Chem. C. 5 (2017) 2488-2503. https://doi.org/10.1039/C7TC00140A

[36] G.R. Berdiyorov, Effect of surface functionalization on the electronic transport properties of Ti3C2 MXene, Epl. 111 (2015) 67002. https://doi.org/10.1209/0295-5075/111/67002

[37] M. Ghidiu, M.R. Lukatskaya, M.Q. Zhao, Y. Gogotsi, M.W. Barsoum, Conductive two-dimensional titanium carbide "clay" with high volumetric capacitance, Nature. 516 (2015) 78-81. https://doi.org/10.1038/nature13970

[38] F. Liu, J. Zhou, S. Wang, B. Wang, C. Shen, L. Wang, Q. Hu, Q. Huang, A. Zhou, Preparation of high-purity V2C MXene and electrochemical properties as li-ion batteries , J. Electrochem. Soc. 164 (2017) A709-A713. https://doi.org/10.1149/2.0641704jes

[39] F. Liu, A. Zhou, J. Chen, J. Jia, W. Zhou, L. Wang, Q. Hu, Preparation of Ti3C2 and Ti2C MXenes by fluoride salts etching and methane adsorptive properties, Appl. Surf. Sci. 416 (2017) 781-789. https://doi.org/10.1016/j.apsusc.2017.04.239

[40] J. Halim, M.R. Lukatskaya, K.M. Cook, J. Lu, C.R. Smith, L.Å. Näslund, S.J. May, L. Hultman, Y. Gogotsi, P. Eklund, M.W. Barsoum, Transparent conductive two-dimensional titanium carbide epitaxial thin films, Chem. Mater. 26 (2014) 2374-2381. https://doi.org/10.1021/cm500641a

[41] M. Serhan, M. Sprowls, D. Jackemeyer, M. Long, I.D. Perez, W. Maret, N. Tao, E. Forzani, Total iron measurement in human serum with a smartphone, AIChE Annu. Meet. Conf. Proc. 2019. https://doi.org/10.1109/JTEHM.2020.3005308

[42] J. Xuan, Z. Wang, Y. Chen, D. Liang, L. Cheng, X. Yang, Z. Liu, R. Ma, T. Sasaki, F. Geng, Organic-base-driven intercalation and delamination for the production of functionalized titanium carbide nanosheets with superior photothermal therapeutic performance, Angew. Chemie. 128 (2016) 14789-14794. https://doi.org/10.1002/ange.201606643

[43] S. Yang, P. Zhang, F. Wang, A.G. Ricciardulli, M.R. Lohe, P.W.M. Blom, X. Feng, Fluoride-free synthesis of two-dimensional titanium carbide (MXene) using a binary aqueous system, Angew. Chemie. 130 (2018) 15717-15721. https://doi.org/10.1002/ange.201809662

[44] W. Sun, S.A. Shah, Y. Chen, Z. Tan, H. Gao, T. Habib, M. Radovic, M.J. Green, Electrochemical etching of Ti2AlC to Ti2CT:X (MXene) in low-concentration

hydrochloric acid solution, J. Mater. Chem. A. 5 (2017) 21663-21668. https://doi.org/10.1039/C7TA05574A

[45] M. Li, J. Lu, K. Luo, Y. Li, K. Chang, K. Chen, J. Zhou, J. Rosen, L. Hultman, P. Eklund, P.O.Å. Persson, S. Du, Z. Chai, Z. Huang, Q. Huang, Element replacement approach by reaction with lewis acidic molten salts to synthesize nanolaminated MAX Phases and MXenes, J. Am. Chem. Soc. 141 (2019) 4730-4737. https://doi.org/10.1021/jacs.9b00574

[46] L. Wang, H. Zhang, B. Wang, C. Shen, C. Zhang, Q. Hu, A. Zhou, B. Liu, Synthesis and electrochemical performance of Ti3C2Tx with hydrothermal process, Electron. Mater. Lett. 12 (2016) 702-710. https://doi.org/10.1007/s13391-016-6088-z

[47] T. Li, L. Yao, Q. Liu, J. Gu, R. Luo, J. Li, X. Yan, W. Wang, P. Liu, B. Chen, W. Zhang, W. Abbas, R. Naz, D. Zhang, Fluorine-free synthesis of high-purity Ti3C2Tx (T=OH, O) via alkali treatment, Angew. Chemie - Int. Ed. 57 (2018) 6115-6119. https://doi.org/10.1002/anie.201800887

[48] Q. Xue, H. Zhang, M. Zhu, Z. Pei, H. Li, Z. Wang, Y. Huang, Y. Huang, Q. Deng, J. Zhou, S. Du, Q. Huang, C. Zhi, Photoluminescent Ti3C2 MXene quantum dots for multicolor cellular imaging, Adv. Mater. 29 (2017) 1-6. https://doi.org/10.1002/adma.201604847

[49] B.C. Wyatt, A. Rosenkranz, B. Anasori, 2D MXenes: Tunable mechanical and tribological properties, Adv. Mater. 2007973 (2021) 1-15. https://doi.org/10.1002/adma.202007973

[50] T. Review, O. Access, Physical properties of 2D MXenes : From a theoretical perspective, J. Phys. : Mater. 3 (2021) 032006. https://doi.org/10.1088/2515-7639/ab97ee

[51] M. Kurtoglu, M. Naguib, Y. Gogotsi, M.W. Barsoum, First principles study of two-dimensional early transition metal carbides, MRS Commun. 2 (2012) 133-137. https://doi.org/10.1557/mrc.2012.25

[52] M. Boota, B. Anasori, C. Voigt, M.Q. Zhao, M.W. Barsoum, Y. Gogotsi, Pseudocapacitive electrodes produced by oxidant-free polymerization of pyrrole between the layers of 2D titanium carbide (MXene), Adv. Mater. 28 (2016) 1517-1522. https://doi.org/10.1002/adma.201504705

[53] S. Nam, S. Umrao, S. Oh, I. Oh, 2D layered Ti3C2Tx negative electrode-based activated carbon woven, 32 (2019) 296-300.

[54] N.N. Wang, H. Wang, Y.Y. Wang, Y.H. Wei, J.Y. Si, A.C.Y. Yuen, J.S. Xie, B. Yu, S.E. Zhu, H.D. Lu, W. Yang, Q.N. Chan, G.H. Yeoh, Robust, lightweight, hydrophobic, and fire-retarded polyimide/MXene aerogels for effective oil/water separation, ACS Appl. Mater. Interfaces. 11 (2019) 40512-40523. https://doi.org/10.1021/acsami.9b14265

[55] Y. Ibrahim, A. Mohamed, A.M. Abdelgawad, K. Eid, A.M. Abdullah, A. Elzatahry, The recent advances in the mechanical properties of self-standing two-dimensional MXene-based nanostructures: Deep insights into the supercapacitor, Nanomaterials. 10 (2020) 1-27. https://doi.org/10.3390/nano10101916

[56] B. Anasori, M.R. Lukatskaya, Y. Gogotsi, 2D metal carbides and nitrides (MXenes) for energy storage, Nat. Rev. Mater. 2 (2017). https://doi.org/10.1038/natrevmats.2016.98

[57] M. Khazaei, M. Arai, T. Sasaki, C.Y. Chung, N.S. Venkataramanan, M. Estili, Y. Sakka, Y. Kawazoe, Novel electronic and magnetic properties of two-dimensional transition metal carbides and nitrides, Adv. Funct. Mater. 23 (2013) 2185-2192. https://doi.org/10.1002/adfm.201202502

[58] J.A. Kumar, P. Prakash, T. Krithiga, D.J. Amarnath, J. Premkumar, N. Rajamohan, Y. Vasseghian, P. Saravanan, M. Rajasimman, Methods of synthesis, characteristics, and environmental applications of MXene: A comprehensive review, Chemosphere. 286 (2022) 131607. https://doi.org/10.1016/j.chemosphere.2021.131607

[59] C.J. Zhang, B. Anasori, A. Seral-Ascaso, S.H. Park, N. McEvoy, A. Shmeliov, G.S. Duesberg, J.N. Coleman, Y. Gogotsi, V. Nicolosi, Transparent, flexible, and conductive 2D titanium carbide (MXene) films with high volumetric capacitance, Adv. Mater. 29 (2017) 1-9. https://doi.org/10.1002/adma.201702678

[60] K. Hantanasirisakul, M.Q. Zhao, P. Urbankowski, J. Halim, B. Anasori, S. Kota, C.E. Ren, M.W. Barsoum, Y. Gogotsi, Fabrication of Ti3C2Tx MXene transparent thin films with tunable optoelectronic properties, Adv. Electron. Mater. 2 (2016) 1-7. https://doi.org/10.1002/aelm.201600050

[61] M. Khazaei, A. Ranjbar, M. Arai, S. Yunoki, Topological insulators in the ordered double transition metals M2′M″C2 MXenes (M′= Mo, W; M″=Ti, Zr, Hf), Phys. Rev. B. 94 (2016) 1-9. https://doi.org/10.1103/PhysRevB.94.125152

[62] Y. Liang, M. Khazaei, A. Ranjbar, M. Arai, S. Yunoki, Y. Kawazoe, H. Weng, Z. Fang, Theoretical prediction of two-dimensional functionalized MXene nitrides as topological insulators, Phys. Rev. B. 96 (2017) 1-9. https://doi.org/10.1103/PhysRevB.96.195414

[63] C. Si, K.H. Jin, J. Zhou, Z. Sun, F. Liu, Large-gap quantum spin hall state in MXenes: D-band topological order in a triangular lattice, Nano Lett. 16 (2016) 6584-6591. https://doi.org/10.1021/acs.nanolett.6b03118

[64] X. Gao, Z. Jia, B. Wang, X. Wu, T. Sun, X. Liu, Q. Chi, G. Wu, Synthesis of NiCo-LDH/MXene hybrids with abundant heterojunction surfaces as a lightweight electromagnetic wave absorber, Chem. Eng. J. 419 (2021) 130019. https://doi.org/10.1016/j.cej.2021.130019

[65] H. An, T. Habib, S. Shah, H. Gao, M. Radovic, Surface-agnostic highly stretchable and bendable conductive MXene multilayers, Sci. Adv. 4 (2018) 1-9. https://doi.org/10.1126/sciadv.aaq0118

[66] J. Zhou, X.H. Zha, M. Yildizhan, P. Eklund, J. Xue, M. Liao, P.O.Å. Persson, S. Du, Q. Huang, Two-dimensional hydroxyl-functionalized and carbon-deficient scandium carbide, ScCxOH, a direct band gap semiconductor, ACS Nano. 13 (2019) 1195-1203. https://doi.org/10.1021/acsnano.8b06279

[67] Z. Kang, Y. Ma, X. Tan, M. Zhu, Z. Zheng, N. Liu, L. Li, Z. Zou, X. Jiang, T. Zhai, Y. Gao, MXene-silicon van der Waals heterostructures for high-speed self-driven photodetectors, Adv. Electron. Mater. 3 (2017) 1-7. https://doi.org/10.1002/aelm.201700165

[68] Z. Kang, Z. Zheng, H. Wei, Z. Zhang, X. Tan, L. Xiong, T. Zhai, Y. Gao, Controlled growth of an Mo2C-graphene hybrid film as an electrode in self-powered two-sided Mo2C-graphene/Sb2S0.42Se2.58 /TiO2 photodetectors, Sensors (Switzerland). 19 (2019) 1-11. https://doi.org/10.3390/s19051099

[69] M. Li, H. Wang, X. Wang, Q. Lu, H. Li, Y. Zhang, S. Yao, Ti3C2/Cu2O heterostructure based signal-off photoelectrochemical sensor for high sensitivity detection of glucose, Biosens. Bioelectron. 142 (2019) 111535. https://doi.org/10.1016/j.bios.2019.111535

[70] Y. Yang, J. Jeon, J.H. Park, M.S. Jeong, B.H. Lee, E. Hwang, S. Lee, Plasmonic transition metal carbide electrodes for high-performance InSe photodetectors, ACS Nano. 13 (2019) 8804-8810. https://doi.org/10.1021/acsnano.9b01941

[71] M. Naguib, J. Halim, J. Lu, K.M. Cook, L. Hultman, Y. Gogotsi, M.W. Barsoum, New two-dimensional niobium and vanadium carbides as promising materials for li-ion batteries, J. Am. Chem. Soc. 135 (2013) 15966-15969. https://doi.org/10.1021/ja405735d

[72] C. Peng, P. Wei, X. Chen, Y. Zhang, F. Zhu, Y. Cao, H. Wang, H. Yu, F. Peng, A hydrothermal etching route to synthesis of 2D MXene (Ti3C2, Nb2C): Enhanced exfoliation and improved adsorption performance, Ceram. Int. 44 (2018) 18886-18893. https://doi.org/10.1016/j.ceramint.2018.07.124

[73] M.H. Tran, T. Schäfer, A. Shahraei, M. Dürrschnabel, L.M. Luna, U.I. Kramm, C.S. Birkel, Adding a new member to the MXene family: Synthesis, structure, and electrocatalytic activity for the hydrogen evolution reaction of V4C3Tx, ACS Appl. Energy Mater. 1 (2018) 3908-3914. https://doi.org/10.1021/acsaem.8b00652

[74] F. Chang, C. Li, J. Yang, H. Tang, M. Xue, Synthesis of a new graphene-like transition metal carbide by de-intercalating Ti3AlC2, Mater. Lett. 109 (2013) 295-298. https://doi.org/10.1016/j.matlet.2013.07.102

[75] E. Kayali, A. Vahidmohammadi, J. Orangi, M. Beidaghi, Controlling the dimensions of 2D MXenes for ultrahigh-rate pseudocapacitive energy storage, ACS Appl. Mater. Interfaces. 10 (2018) 25949-25954. https://doi.org/10.1021/acsami.8b07397

[76] H. Kim, B. Anasori, Y. Gogotsi, H.N. Alshareef, Thermoelectric properties of two-dimensional molybdenum-based MXenes, Chem. Mater. 29 (2017) 6472-6479. https://doi.org/10.1021/acs.chemmater.7b02056

[77] J. Yang, M. Naguib, M. Ghidiu, L.M. Pan, J. Gu, J. Nanda, J. Halim, Y. Gogotsi, M.W. Barsoum, Two-dimensional Nb-based M4C3 solid solutions (MXenes), J. Am. Ceram. Soc. 99 (2016) 660-666. https://doi.org/10.1111/jace.13922

[78] A. Lipatov, M. Alhabeb, M.R. Lukatskaya, A. Boson, Y. Gogotsi, A. Sinitskii, Effect of synthesis on quality, electronic properties and environmental stability of individual monolayer Ti3C2 MXene flakes, Adv. Electron. Mater. 2 (2016). https://doi.org/10.1002/aelm.201600255

[79] J. Yan, C.E. Ren, K. Maleski, C.B. Hatter, B. Anasori, P. Urbankowski, A. Sarycheva, Y. Gogotsi, Flexible MXene/graphene films for ultrafast supercapacitors with outstanding volumetric capacitance, Adv. Funct. Mater. 27 (2017) 1-10. https://doi.org/10.1002/adfm.201701264

[80] X. Zhang, Y. Liu, S. Dong, J. Yang, X. Liu, Surface modified MXene film as a flexible electrode with ultrahigh volumetric capacitance, Electrochim. Acta. 294 (2019) 233-239. https://doi.org/10.1016/j.electacta.2018.10.096

[81] K. Maleski, C.E. Ren, M.Q. Zhao, B. Anasori, Y. Gogotsi, Size-dependent physical and electrochemical properties of two-dimensional MXene flakes, ACS Appl. Mater. Interfaces. 10 (2018) 24491-24498. https://doi.org/10.1021/acsami.8b04662

[82] J. Zhang, N. Kong, S. Uzun, A. Levitt, S. Seyedin, P.A. Lynch, S. Qin, M. Han, W. Yang, J. Liu, X. Wang, Y. Gogotsi, J.M. Razal, Scalable manufacturing of free-standing, strong Ti3C2Tx MXene films with outstanding conductivity, Adv. Mater. 32 (2020) 1-9. https://doi.org/10.1002/adma.202001093

[83] G. Ying, A.D. Dillon, A.T. Fafarman, M.W. Barsoum, Transparent, conductive solution processed spincast 2D Ti2CTx (MXene) films, Mater. Res. Lett. 5 (2017) 391-398. https://doi.org/10.1080/21663831.2017.1296043

[84] F. Du, H. Tang, L. Pan, T. Zhang, H. Lu, J. Xiong, J. Yang, C. Zhang, Environmental friendly scalable production of colloidal 2D titanium carbonitride MXene with minimized nanosheets restacking for excellent cycle life lithium-ion batteries, Electrochim. Acta. 235 (2017) 690-699. https://doi.org/10.1016/j.electacta.2017.03.153

[85] J. Jeon, H. Choi, S. Choi, J.H. Park, B.H. Lee, E. Hwang, S. Lee, Transition-metal-carbide (Mo2C) multiperiod gratings for realization of high-sensitivity and broad-spectrum photodetection, Adv. Funct. Mater. 29 (2019) 1-7. https://doi.org/10.1002/adfm.201905384

[86] K. Montazeri, M. Currie, L. Verger, P. Dianat, M.W. Barsoum, B. Nabet, Beyond gold: Spin-coated Ti3C2-based MXene photodetectors, Adv. Mater. 31 (2019) 1-9. https://doi.org/10.1002/adma.201903271

Recent Advances and Allied Applications of MXenes
Materials Research Foundations 155 (2024) 103-114

Materials Research Forum LLC
https://doi.org/10.21741/9781644902875-5

Chapter 5

Applications of MXenes in Electrocatalysis

Keerthiga Gopalram*

Chemical Engineering Department,   SRM Institute of Science and Technology (SRMIST),
Kattankulathur, Chengalpattu -603 203, Tamil Nadu, INDIA

Abstract

MXenes belong to the group of two-dimensional inorganic compounds and have been recently exploited to solve major research problems. The doping, etching, and composite preparation with MXene will lead to synergistic enhancement of active sites. This book chapter summarizes recent updates on MXene as an electrocatalyst for reactions of hydrogen evolution, oxygen reduction, batteries, supercapacitor, nitrogen mitigation, and CO_2 mitigation reaction. The adaptability of MXenes with its modification of surface terminal groups facilitates MXenes for specific electro-catalytic reactions.

Keywords

MXene, Terminal Groups, Reactions for Hydrogen Evolution, Reactions for Oxygen Evolution, Electro Catalyst, Carbon Dioxide Mitigation Reduction

Contents

1.   Introduction

Electrocatalyst acts as an agent to overcome the thermodynamic energy barrier for any specific electrode reaction, facilitating the chemical interaction on the electrode surface. Though there are widely used metals, metal oxides, hybrid materials, and supports, the specific attention of electrocatalysts for their activity, stability, and selectivity is the need of the hour. MXene is one such material that can be modified to suit any specific electrochemical reaction. The search for new materials to solve energy and environmental problems demands the innovation of MXene-based two-dimensional materials as a catalyst, support, and active material. The development of heterogeneous clay-inspired MXenes by intercalation, de-alumination, functionalization, hybridization, and its applications in various fields has been discussed briefly. The electro-catalytic applications of MXene for reactions of hydrogen and oxygen evolution reaction, sensing, carbon dioxide and nitrogen mitigation reaction, and environmental remediation has been covered in this chapter. The adaptive version of MXene with its tuned hydrophilic and hydrophobic nature makes it a versatile catalyst for diverse applications. The futuristic scope of MXene and the literature gap in certain fields helps to gain momentum to reap the best for commercial applications.

MXene where "M" represents d block elements, "X" represents heteroatom, and T_x represents functional groups in which the chemical formula and the thickness can be controlled by the number of n layers with n=1-3 and chemically represented by $M_{n+1}X_nT_x$, [2]. Fig. 1 depicts its SEM image and processing of MXene nanosheets. In the MAX phase, the center "A" element can be remodified with any transition metal carbonitrides, metal carbides, and nitrides.

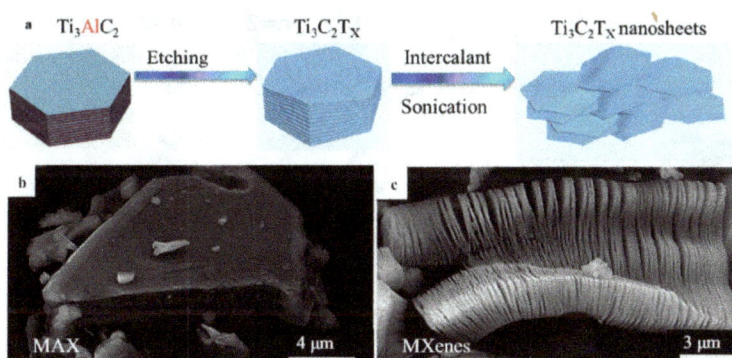

Figure 1: Illustration of etching and SEM image of MAX and MXenes, reproduced from Bai et al., 2021 [1]

MXene ($M_{n+1}X_n$) has been represented as a hexagonal close-packed structure. They find applications in supercapacitors, and batteries, as they possess long interlayer distances, good hydrophilicity, and appreciable conductivity [3]. MXene has a negatively charged surface compared to charge-neutral graphene due to the presence of enhanced functional groups of -OH, -O, and –F. Ternary transition Metals with boron atoms such as MBenes such as Mo_2B_2, Fe_2B, MnB, Nb_5B_2, and Ta_3B_4 show diverse structures and functions and be used for various applications [4]. Currently, MXenes find applications in the field of optics, thermoelectric devices, sensors, variety of chemical reactions, supercapacitors, water splitting for hydrogen evolution, Li-ion batteries, and other catalytic processes.

1.1 Features of MXene as an Electrocatalyst

The distinctive features of MOF are not restricted to good electrical conductivity, accumulated surface area, hydrophilic and hydrophobic character based on terminal supporting groups, better stability for the targeted molecule for specific reaction [4]. Modification in MXene through the surface terminal groups of –H coordination helps to stabilize the intermediates, [5], Fig. 2 shows terminal groups of MXene and its modification.

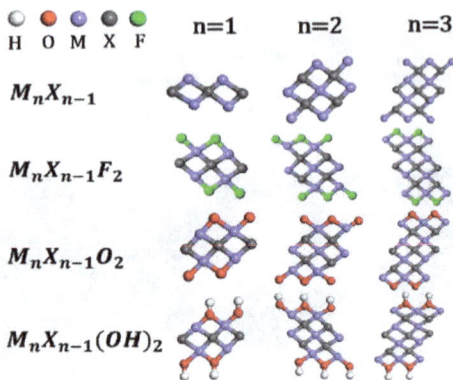

Figure 2: MXene with different terminal oxygen separation [3]

MXene can be accounted as the possible non-carbon support and favors to be used under harsh conditions [6]. Metallic Ti in MXene offers high durability when tested for oxidative and reducing conditions where carbon in the MXene shows high conductive channel favoring reduction [7,8].

1.2 Mechanical properties of MXENE

MXenes may have metallic characteristics similar to graphene of comparable thickness. Polyvinyl alcohol can be hybridized with MXene to form Ti_3C_2Tx/polymer composite films and has excellent flexibility, tensile strength, and mechanical properties [9].

1.3 Electrical structures of MXenes

MXene includes a carbon layer with characteristics comparable to graphene. Meanwhile, because it contains a transition metal layer, with similar properties of transition metal oxide. It has strong electrical conductivity, and good energy storage. Furthermore, the electron is confined in the ultrathin area, demonstrating exceptional conductivity, anisotropic carrier mobility, ultrathin structure, and hexagonal lattice symmetry. The electrical characteristics of MXenes appear to be affected by the functional groups. It has been found that virgin MXenes (e.g., Ti_3C_2) have metallic properties, but MXenes exhibit semiconductor properties with addition of hetero atoms (F, O) and hydroxyl groups [10].

2. Synthesis of MXenes

Exfoliation of chemical bonds of "M" and "A" phase poses challenges and hence demands chemical etching. The most common way is to selectively erode A phase with retained A and X phase undisturbed in the structure (Fig. 3).

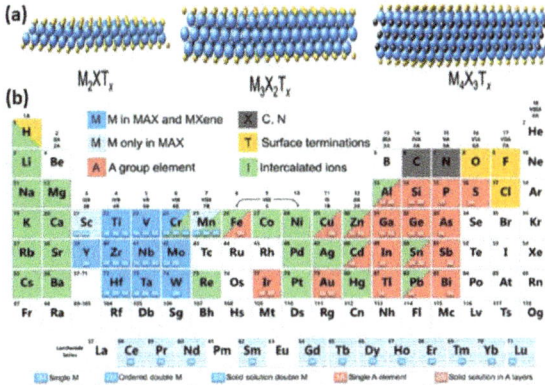

Figure 3: Representation of MXenes, Periodic table showing MXene precursors (blue) and red represents MXene what can be potentially selective to etch to form MXene [2].

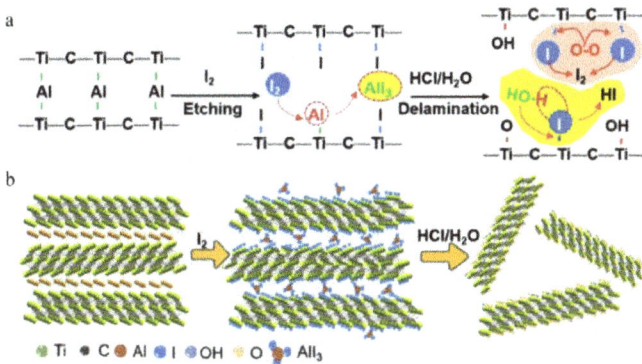

Figure 4: Iodine assisted etching of 2D MXenes, reproduced from Bai et al., 2020 [1]

The substitution of appropriate groups in the lattice induces hydrophobicity (-F terminations) and hydrophilicity (–OH terminations). Fig. 4 shows iodine assistant etching of MXene. The Ti_3C_2Tx MXene, for example, can be made by selectively removing "A" from its parent MAX phase, Ti_3AlC_2. The MXenes are made up of one-step chemical method while other methods of MXene modification are discussed elsewhere.

3. Applications of MXene as electrocatalyst

3.1 MXene for hydrogen evolution reaction

The properties of MXene can be improved by combining with other metals, doping with metal and nonmetals, inducing the change in surface termination groups. The MXene can be doped with double layered hydroxides, chalcogenides and noble metals which improves the HER performance. Formation of composites can be made by self-assembly method and bottom to top growth method. The preferable chemical reaction induced changes are accommodated in the bottom-up growth method and the properties of interactive forces play a role in the electrostatic attraction method. The modification in the 2 D morphologies promotes hierarchical structures that exhibit versatile HER properties as electro catalysts where Table 1 illustrates few literature updates.

Versatile HER catalyst suffers from cost, availability but is active for HER under various operating conditions. Dispersion of Pt on appropriate support has been widely practiced where MXene with its features of good conductivity, better hydrophilicity and high surface area helps in dispersion of Pt based HER catalyst and prevents agglomeration of catalyst. Alloying metal with non-precious metals also improves the utilization of Platinum.

The modification of Ti_3C_2 MXene by oxygen termination enhances the HER activity compared to the –F induced modification. In addition, polyaniline and palladium metal doped MXene PANI/Pd/MXene nanocomposite show better oxidative properties for methanol with enhanced peak current density than Pd/MXene. The catalyst was stable up to 100 cycles with enhanced methanol oxidation reaction and up to three times higher current density [11].

Understanding derived from HER application of MXene are improvement of stability of MXENE by forming oxidative resistive layer, development of large-scale application of MXene, development of 3 D architects of MXene for dispersing active materials in the support, activation of terminal groups of MXene for HER electro catalyst [3]. The performance of catalysts with –S, -Cl and –Br induced catalysts needs investigation. The doping with nitrogen with MXene results in improved electrical conductivity and reactivity properties which offers higher durability and low overpotential for HER, with decreased Gibbs free energy [5].

DFT simulation aided to design Mo_2CTx and Ti_2CTx for HER. The catalysts have been compared without support where basal planes of Mo_2CTx were more active than bare MoS_2 [12].

3.2   MXene for nitrogen reduction reaction

The choice of MXene for nitrogen fixation is due to its interaction of terminal oxygen on MXenes to hydrogen evolution and carbon dioxide fixation followed by weakening of the bonds for nitrogen reduction rather than CO_2 reduction as proved by theoretical simulation.

The conversion of nitrogen to useful chemicals remains a distant dream due to lack of appropriate catalysts and good technology. The N_2 capture study on MXene showed V_3C_2 and Nb_3C_2 as a good catalyst at lower reaction energies. In another DFT study, the Gibbs free energy for N_2 adsorption has been studied on a series of metal catalysts onto $Ti_3C_2O_2$ for lower reaction energies.

The N_2 adsorption is favoured on central Ti atom of MXene Ti_3C_2Tx than compared to other sites which favours reduction of N_2. F-free Ti_3C_2Tx nanosheets when studied for N_2 reduction forms ammonia with good yield and Faradaic efficiency (9.1%) [13]. Hydrothermally prepared $TiO_2/$ Ti_3C_2Tx hybrids when studied for N_2RR showed 8.42 % FE at -0.6 V vs RHE. MnO_2-decorated Ti_3C_2Tx MXene has also been reported for its good active sites, adsorption and activation of N_2. Similarly, M_2NO_2 types MXenes have been doped in single transition metal atoms using DFT studies for Nitrogen reduction reaction. DFT studies infer the intermediate formation of *NNH and *NH_2 converges into NH_3 ammonia formation.

3.2   MXene for carbon dioxide reduction reaction

The specific advantage of MXene for CO_2 reduction reaction is conductive to efficient gas phase transfer, availability of metal at its outskirts for interaction with CO_2, stability of MXene in aqueous medium, surface hydrophilic structure for imbibed reaction with target species.

Ti_3C_2 quantum dots with Cu metal doped on Cu mesh fabricated by electrostatic self-assembly method shows promising results for methanol formation (8.25 times higher than Cu_2O nanowires). Promising initial results of MXene for mitigation of carbon dioxide needs to broadcast more research investigation, where few results have been tabulated in Table 1.

Table 1: Summary of characteristics of MXene doped catalyst for electrochemical study.

Catalyst	Specific activity towards reaction	References
MXenes for OER ACTIVITY	**Over potential**	
N–$Ti_3C_2T_x$-35 in H_2SO_4	162 mV;	3
N–$Ti_3C_2T_x$@600 in H_2SO_4	198 mV;	3
Nb_4C_3T in 1.0 M KOH	398 mV;	3
$Ti_3C_2T_x$-FeOOH in 0.1 M KOH	420 mV	3
Ti_3C_2/g-C_3N_4 in 0.1 M KOH	420 mV	3
MXenes for HER activity	**Over potential and Tafel slope (mV dec^{-1})**	**References**
Ti_2CT_x in H_2SO_4	609 mV	14
Flakes of Ti_3C_2 in H_2SO_4	385 mV	14
MoS_2@Mo_2CT_x in KOH	207 mV	1
$Ti_3C_2T_x$ flakes in H_2SO_4	385 mV	1
$Ti_3C_2T_x$- 450 in H_2SO_4	266 mV at 109.8 mV dec^{-1}	1
$VeTi_4N_3T$ in H_2SO_4	330 V at 107 mV dec^{-1}	13
$CoTi_3C_2$ in 1 M KOH	480 V at 147 mV dec^{-1}	13
MXenes for OER activity	**Over potential mV, Tafel slope (mV dec^{-1})**	**References**
$Ti_3C_2T_x$-CoBDC 1 M KOH	410 V at 48.2 mV dec^{-1}	14
Fe-N-C/MXene in 0.1 KOH	0.84 V	14
FeOOH NSs/Ti_3C_2 1 M KOH	400 V at 95 mV dec^{-1}	14
Ti_3C_2Tx/g-C_3N_4 1 M KOH	420 V at 74.6 mV dec^{-1}	14
Ti_3C_2Tx-FeOOH 1 M KOH	420 V at 31.7 mV dec^{-1}	14
Ti_3C_2Tx/g-C_3N_4 1 M KOH	420 V at 74.6 mV dec^{-1}	13
MXenes for batteries (method of synthesis)	**Specific capacitance**	**References**
$Ti_3C_2T_x$/Ag by Direct reduction	310 mA hg-1 (5000 cycles)	14
PDDA-NPCN/Ti_3C_2	583.7 mA h g^{-1} at 0.1 A g^{-1}	14
Sb doped MXene	548 mA h g^{-1} at 50 mA g^{-1}	14
Si/$Ti_3C_2T_x$	1067.6 mA h g^{-1} at 300 mA g^{-1}	14

Transition Metal doped MXene has been studied for CO_2 storage and mitigation by electro and photo electrochemical methods. The reason for appraisal of M_2C MXene is not restricted to the following, (i) availability of broad terminated surfaces for CO_2 capturing, (ii) metallic character favoring photo and electrochemical reactions, (iii) selective absorption of CO_2 over absorption of H_2O.

Methane formation from CO_2 has been reported on Cr_3C_2 and Mo_3C_2 doped MXenes due to the formation of spontaneous intermediates of $OCHO^.$ and HOCO from CO_2 at low overvoltage of 1.05, 1.3 eV respectively. W_2CO_2 and Ti_2CO_2 has also reported to favor

CO_2 to formic acid by *HCOOH pathway rather than *CO pathway and has been proved as a good catalyst for CO_2RR.

High conductive property of MXene and OH activation of MXene promotes activation and reduction of CO_2 molecules. Ti_3CO_2 and W_2CO_2 are the O- terminated MXenes when studied for electrochemical reduction of CO_2 promoting formic acid at low overpotential and good selectivity. Along these lines, MXene can be processed for multi electron reduction targeting methane or ethylene as a product. The areas of electrochemical reduction of CO_2 awaits the experimental results to support the theoretical inference.

3.4   MXene for environmental remediation

Surface tuned MXene has also been studied for sensing various environmental pollutants and water purification, heavy metal ions removal from nuclear wastes (Re (IV), Th (IV), and Eu (III) and also as a chemical adsorbent for radioactive ions.

MXene also finds applications in environmental remediation for the good adsorption of methylene blue when combined with NaOH- $Ti_3C_2R_x$ complex of MXene. MXene with polymer coupling has been studied for dye degradation and MXene with phytic acid (PA) has been reported for degradation of textile dyes such as methylene blue and Rhodamine B (RhB) dyes. Further, its feasibility for commercial applications needs to be verified and accessibility of catalyst in sunlight awaits more experimental validation.

3.5   MXene-based electrocatalysts for ORR

In general, carbon in Ti ensures conductivity for charge transfer and Ti ensures corrosive resistance property of Ti where TiC nanoparticles can be compared with Ti in the MXene layer. Though Platinum a versatile catalyst for oxygen evolution reaction, the dispersion of Pt nanoparticles by impregnation in various forms of Pt $(Cl_4)_2$, Pt $(Cl_6)_4$, Pd $(Cl_4)_2$ etc. has been poorly dispersed on Ti_3C_2Tx support. Though the addition of surface termination brings changes in the property, the terminal of –F ions induces hydrophobicity which brings in the repulsive properties of Pt nanoparticles. To improve the dispersion of metal precursors on the support, poly (diallyl dimethylammonium chloride) (PDDA) functionalization has been attempted to improve the surface charge properties and also to promote and regulate the shape of nanoparticles on the support. In addition, PDDA also prevents restacking of Ti_3C_2Tx during the synthesis process.

Impregnation of TiO_2 with Pt supported MXene leads to improved semiconducting properties of TiO_2 with light induced charge transfer, better illumination and active molecules activation. The electrochemical performance for oxygen evolution reaction of carbon supported MXene was higher than compared to CNT supported catalyst. The modification of MXene as support shows good output for oxygen reduction reactions from literature as shown in Table 1 and shows promising results for commercialization.

3.6   MXene for batteries storage and supercapacitors

The property of MXene can also be improved by addition of metals, monoatomic and 2 D materials while helping in dispersion of active metals and improve the charge carrying properties of MXene. The ways and means to improve the properties of MXene involves modification of functional groups present at the surface and terminal metal sites. Fig 5a and Fig 5b show the method of formation of PVP-Sn (IV) doped MXene and the cyclic stability of the material for battery capacity. The inclusion of MXene in the catalyst promotes energy capacity, specific energy density and power density, if doped with suitable metals.

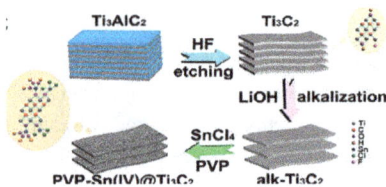

Figure 5a: Schematic illustration of synthesis of PVP-Sn(IV)@Ti₃C₂, reproduced from Li and Bang, 2021 [10].

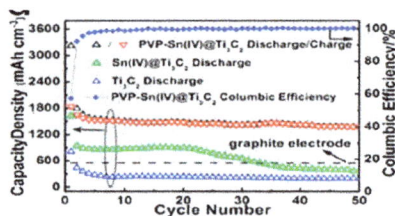

Figure 5b: Cyclic stability of PVP-Sn(IV)@Ti₃C₂

Conclusion

Futuristic scenario will focus on extending the catalytic application of MXene for various applications of CO_2 reduction, N_2 reduction, Soot combustion and photo and photoelectro based environmental mitigation.   Improving the catalytic activity for any desired application will enhance the activity and yield for the desired reaction.

The reason for choice of MXene as electro catalyst may not be restricted to the following properties of significant metal conductivity, the terminal groups have multi valence electrons with good redox property which are more active than other supports of graphene, good stability in the aqueous medium, and their ends with hydrophilic surface improve the interaction with aqueous molecule for any targeted applications. Inferences from electro catalytic application of MXene are promising in terms of experimental and theoretical approach, however, more experiments in diverse environments are needed to harness the versatile properties of MXENE.

Acknowledgments

GK acknowledges the Start-up **Research Grant (SRG)** of DST-SERB with File no: DST/SERB/SRG/001396/ES for its funding and establishment of facilities at SRMIST.

References

[1] S.G. Peera, C. Liu, A.K. Sahu, M. Selvaraj, M.C. Rao, T.G. Lee, R. Koutavarapu, J. Shim, and L. Singh, Recent advances on MXene-based electrocatalysts toward oxygen reduction reaction: A focused review, Adv. Mater. Interfaces 8 (2021) 2100975. https://doi.org/10.1002/admi.202100975

[2] S.G. Peera, C. Liu, J. Shim, A.K. Sahu, T.G. Lee, M. Selvaraj, R. Koutavarapu, MXene (Ti3C2Tx) supported electrocatalysts for methanol and ethanol electrooxidation: A review, Ceram. Int. 47 (2021) 28106-28121. https://doi.org/10.1016/j.ceramint.2021.07.075

[3] S. Bai, M. Yang, J. Jiang, X. He, J. Zou, Z. Xiong, G. Liao and S. Liu, Recent advances of MXenes as electrocatalysts for hydrogen evolution reaction, NPJ 2D Mater. Appl. 5 (2021) 78. https://doi.org/10.1038/s41699-021-00259-4

[4] M. Elancheziyan, M. Eswaran, C.E. Shuck, S. Senthilkumar, S. Elumalai, R. Dhanusuraman, V.K. Ponnusamy, Facile synthesis of polyaniline/titanium carbide (MXene) nanosheets/ palladium nanocomposite for efficient electrocatalytic oxidation of methanol for fuel cell application, Fuel 303 (2021) 121329-121338. https://doi.org/10.1016/j.fuel.2021.121329

[5] Z. Kang, M.A. Khan, Y. Gong, R. Javed, Y. Xu, D. Ye, H. Zhao and J. Zhang, Recent progress of MXenes and MXene-based nanomaterials for the electrocatalytic hydrogen evolution reaction, J. Mater. Chem. A 9 (2021) 6089-6108. https://doi.org/10.1039/D0TA11735H

[6] K. Kannan, K.K. Sadasivuni, A.M. Abdullah and B. Kumar, Current trends in MXene-based nanomaterials for energy storage and conversion system: A mini-review, Catalysts 10 (2020) 495-523. https://doi.org/10.3390/catal10050495

[7] H.J. Liu, B. Dong, Recent advances and prospects of MXene-based materials for electrocatalysis and energy storage, Mater. Today Phys. 20 (2021) 100469-100493. https://doi.org/10.1016/j.mtphys.2021.100469

Materials Research Forum LLC
https://doi.org/10.21741/9781644902875-5

[8] J. Liu, W. Peng, Y. Li, F. Zhang, X. Fan, 2D MXene-based materials for electrocatalysis, Trans. Tianjin University, 26 (2020)149-171. https://doi.org/10.1007/s12209-020-00235-x

[9] L.l. Yu, J.Z. Qin, W.J. Zhao, Z.G. Zhang, J. Ke, and B.J. Liu, Advances in two-dimensional MXenes for Nitrogen electrocatalytic reduction to ammonia, Int. J. Photo. Energy 11 (2020) 5251431-5251444. https://doi.org/10.1155/2020/5251431

[10] T.P. Nguyen, D.M.T. Nguyenc, D.L. Trand, H.K. Le, D.V.N. Vof, S.S. Lamg, R.S. Varmah, M. Shokouhimehri, C.C. Nguyenj, Q.V. Lej, MXenes: Applications in electrocatalytic, photocatalytic hydrogen evolution reaction and CO2 reduction, Mol. Catal. 486 (2020) 110850-110868. https://doi.org/10.1016/j.mcat.2020.110850

[11] Z.W. Seh, K.D. Fredrickson, B. Anasori, J. Kibsgaard, A.L. Strickler, M.R. Lukatskaya, Y. Gogotsi, T.F. Jaramillo, and A. Vojvodic, Two-dimensional molybdenum carbide (MXene) as an efficient electrocatalyst for hydrogen evo lution, ACS Energy Lett. 1 (2016) 589−594 https://doi.org/10.1021/acsenergylett.6b00247

[12] M.M. Tunesi, R.A. Soomro, X. Han, Q. Zhu, Y. Wei and B. Xu, Application of MXenes in environmental remediation technologies, Nano Convergence 8 (2021) 5. https://doi.org/10.1186/s40580-021-00255-w

Keyword Index

About the Editors

Dr. Inamuddin is working as an Assistant Professor at the Department of Applied Chemistry, Aligarh Muslim University, Aligarh, India. He obtained a Master of Science degree in Organic Chemistry from Chaudhary Charan Singh (CCS) University, Meerut, India, in 2002. He received his Master of Philosophy and Doctor of Philosophy degrees in Applied Chemistry from Aligarh Muslim University (AMU), India, in 2004 and 2007, respectively. He has extensive research experience in multidisciplinary fields of Analytical Chemistry, Materials Chemistry, and Electrochemistry and, more specifically, Renewable Energy and Environment. He has worked on different research projects as a project fellow and senior research fellow funded by the University Grants Commission (UGC), Government of India, and the Council of Scientific and Industrial Research (CSIR), Government of India. He has received the Fast Track Young Scientist Award from the Department of Science and Technology, India, to work in the area of bending actuators and artificial muscles. He has also received the Sir Syed Young Researcher of the Year Award 2020 from Aligarh Muslim University. He has completed four major research projects sanctioned by the University Grant Commission, Department of Science and Technology, Council of Scientific and Industrial Research, and Council of Science and Technology, India. He has published 210 research articles in international journals of repute and nineteen book chapters in knowledge-based book editions published by renowned international publishers. He has published 180 edited books with Springer (U.K.), Elsevier, Nova Science Publishers, Inc. (U.S.A.), CRC Press Taylor & Francis Asia Pacific, Trans Tech Publications Ltd. (Switzerland), IntechOpen Limited (U.K.), Wiley-Scrivener, (U.S.A.) and Materials Research Forum LLC (U.S.A). He is a member of various journals' editorial boards. He has served as Associate Editor for journals (Environmental Chemistry Letter, Applied Water Science and Euro-Mediterranean Journal for Environmental Integration, Springer-Nature), Frontiers Section Editor (Current Analytical Chemistry, Bentham Science Publishers), Editorial Board Member (Scientific Reports-Nature) and Review Editor (Frontiers in Chemistry, Frontiers, U.K.) He has also guest-edited various special thematic issues for the journals of Elsevier, Bentham Science Publishers, and John Wiley & Sons, Inc. He has attended as well as chaired sessions at various international and national conferences. He has worked as a Postdoctoral Fellow, leading a research team at the Creative Research Initiative Center for Bio-Artificial Muscle, Hanyang University, South Korea, in the field of renewable energy, especially biofuel cells. He has also worked as a Postdoctoral Fellow at the Center of Research Excellence in Renewable Energy, King Fahd University of Petroleum and Minerals, Saudi Arabia, in the field of polymer electrolyte membrane fuel cells and computational fluid dynamics of polymer electrolyte membrane fuel cells. He is a life member of the Journal of the Indian Chemical Society. His research interest includes ion exchange materials, a sensor for heavy metal ions, biofuel cells, supercapacitors and bending actuators.

Dr. Mohammad A. Jafar Mazumder has been serving as a Professor of Chemistry at King Fahd University of Petroleum & Minerals (KFUPM), Saudi Arabia. He has extensive experience in designing, synthesizing, and characterizing various organic compounds, ionic and thermo-responsive polymers for corrosion, water treatment, and biomedical applications. Dr. Jafar Mazumder obtained his B.Sc (Hons.), M.Sc (Chemistry) from Aligarh Muslim University, India, MS (Chemistry) from KFUPM, Saudi Arabia, and Ph.D. in Chemistry (2009) from McMaster University, Canada.

In more than 20 years of academic research, Dr. Jafar Mazumder has had the opportunity to work with several international collaborative research groups and has exposed himself to a broad range of research areas. Dr. Jafar Mazumder secured 7 US patents, published more than 85 articles in peer-reviewed journals, 37 conference proceedings, 9 book chapters, and co-edited 4 books with Springers and Trans Tech publications. He is awarded as a Fellow of the Royal Society of Chemistry and Chartered Chemist, Association of Chemical Profession of Ontario, Canada. Besides, Dr. Jafar Mazumder is a member of the American Chemical Society (ACS), Canadian Society for Chemistry (CSC), Canadian Biomaterial Society (CBS), and a life member of the Bangladesh Chemical Society (BCS). In his academic career, he was awarded numerous national and international scholarships and awards including the prestigious Indian Council for Cultural Relations (ICCR) Scholarship from Govt. of India for undergraduate studies in India, Aligarh Muslim University undergraduate & graduate Gold medal, and certificate of excellence from the Ministry of Human Resource Development, Govt. of India, and MITACS postdoctoral fellowship (Canada) for pursuing postdoctoral research in Chemical and Biomedical Engineering.

Currently, Dr. Jafar Mazumder is actively involved in several ongoing universities (KFUPM), government (KACST, NSTIP), and client (Saudi Aramco) funded projects in the capacity of principal and co-investigators. His current research interest includes the design, synthesis, and characterization of various modified monomers and polymers for potential use in the inhibition of mild steel corrosion in oil and gas industries and the preparation of multilayered polyelectrolyte coated membranes for the removal of heavy metals and organic contaminants from aqueous water samples. The long-term scientific goal of Dr. Jafar Mazumder is not merely to make science fun and entertaining for people. It is to engage them with a multidisciplinary scientific mission at a deeper level to create a space through which they can interact with scientific ideas, develop connections between science, engineering, and biology, and thoughts of their own to contribute to society. He feels this goal and engaging personality make him a pleasant person to work with and help inspire his co-workers in any professional setting.

Dr. Mohammad Luqman has 12+ years of post-PhD experience in Teaching, Research, and Administration. Currently, he is serving as an Assistant Professor of Chemical Engineering at Taibah University, Saudi Arabia. Before joining here, he served as an Assistant Professor in the College of Applied Science at A'Sharqiyah University, Oman, and in the College of Engineering at King Saud University, Saudi Arabia. He served as a Research Engineer at SAMSUNG Cheil Industries, South Korea. Moreover, he served as a post-doctoral fellow at the Artificial Muscle Research Center, Konkuk University, South Korea, in the field of Ionic Polymer Metal Composites for the development of Artificial Muscles, Robotic Actuators, and Dynamic Sensors. He earned his PhD degree in the field of Ionomers (Ion-containing Polymers), from Chosun University, South Korea. He successfully served as an Editor to three books, published by world-renowned publishers. He published numerous high-quality papers and book chapters. He is serving as an Editor and editorial/review board member to many International SCI and non-SCI journals. He has attracted a few important research grants from industry and academia. His research interests include but are not limited to the Development of Ionomer/Polyelectrolyte/non-ionic Polymer Nanocomposites/Blends for Smart and Industrial/Engineering Applications.

Dr. Mohammad Faraz Ahmer is presently working as an Assistant Professor in the Department of Electrical Engineering, Mewat Engineering College, Nuh Haryana, India, since 2012 after working as a Guest Faculty at University Polytechnic, Aligarh Muslim University Aligarh, India, in 2009-2011. He completed his M.Tech. (2009) and Bachelor of Engineering (2007) degrees in Electrical Engineering from Aligarh Muslim University, Aligarh in the first division. He obtained a Ph.D. degree in 2016 on his thesis entitled "Studies on Electrochemical Capacitor Electrodes". He has published six research papers in reputed scientific journals. He has edited two books with Materials Science Forum, U.S.A. His scientific interests include electrospun nano-composites and supercapacitors. He has presented his work at several conferences. He is actively engaged in searching for new methodologies involving the development of organic composite materials for energy storage systems.

www.ingramcontent.com/pod-product-compliance
Lightning Source LLC
Chambersburg PA
CBHW071711210326
41597CB00017B/2438